The Technics of Flour Milling

.. THE ..

TECHNICS

OF

FLOUR MILLING.

A Handbook for Millers.

BY

WILLIAM HALLIWELL,

Technical Editor "The Miller,"
Registered Teacher of Milling Technology,
Lecturer on Flour Milling, &c.

LONDON:
STRAKER BROTHERS, LIMITED,
"The Bishopsgate Press," 44—47, Bishopsgate Without, E.C.

1904.

TS 2,145
113

Creas
f. a. d. i.

AUTHOR'S PREFACE.

———✦———

A SECOND EDITION of "The Technics of Flour Milling" scarcely needs a second preface, the author being highly gratified at the reception accorded to the initial imprint. It is, however, the usual custom, and, in as few words as may be, the following pages are presented to millers the world over in a spirit of confidence engendered by the success of previous effort. Flour milling to-day is fast approaching an exact science, and it would surely be a pity if nothing were done to keep its rules in evidence. My good friend W. R. Voller has decided not to re-issue his great work on trade technics, and, feeling that some help was required by those who are constantly entering the trade, and also by men of experience in every-day practice, no apology is needed in putting this work before English-speaking millers in all lands. Thanks are hereby tendered to all those who have been so generous in their assistance —to engineers and experts, to millers in many parts of the country, and to the trade journals—in what has been a labour of love.

W. H.

ROMFORD, *January, 1904.*

LIST OF CHAPTERS.

——◆——

INDEX.

THE

TECHNICS OF FLOUR MILLING

CHAPTER I.

HISTORICAL.

THE origin of power and system in flour-milling processes is directly derived from the Greek watermill, a very simple machine with a horizontal wheel but no gearing. It survives in some districts to this day. The great departure from this was when the Roman water-wheel made its appearance. This latter had a vertical wheel and drove the stones through the intervention of cog-wheels. One hundred years intervened between the invention of the Greek mill and the improvement made by the Romans, and after the lapse of another four hundred years the Romans brought out the floating mill or ship mill. This was a structure with a water-wheel in the middle, and the power was obtained from the flowing stream. Taken altogether, watermilling held sway over a thousand years after the Greek period of invention. Then the windmill was sprung upon the world, and caused a tremendous sensation. It was very rapidly adopted. In its first state it

B

was the old tripod mill, then it was improved into a turret, and from the latter was eventually evolved the power mill of to-day. The art of flour milling has been of very slow growth, gradually unfolding its possibilities from absolute crudity to the perfection of the present time very leisurely, as generation succeeded generation. We must not altogether call this a fault, because for hundreds of years the peculiar laws of nearly all countries denied the miller freedom to trade when and where he liked. In fact, up to a rather recent period " soke " mills, as they were called, had to pay the ruling powers a royalty before they were allowed to grind at all. It can easily be understood that these various restrictions, coupled, no doubt, with limited capital, were against personal encouragement. There was not the incentive of competition, either in quality or price, that we are so painfully familiar with to-day, until just about 100 years ago.

To quote Ovid, the first food of man was herbs; then came nuts, and later was discovered the acorn. These man used until the goddess Ceres, having compassion on him, substituted wheat.

Classic literature abounds in allusions to early corn milling by stone grinding; in fact, it is stated on the highest authority that it is the oldest continuously-conducted industry in the world. Most probably the very earliest efforts were confined to simply pounding the nuts, berries, and acorns into a sort of meal. It was only at the time when the Roman Empire was in its glory that the first mechanically-driven mill was invented, and we are told that it excited their wonder and admiration. The milling of wheat by stones is said to date back at least 6,000 years. For 4,000 years nothing but the hand-stone was known. Curiously enough, in nearly all countries it was the woman who had the laborious part of the work to do. Even to-day among the tribes of North American Indians the hand-stone is a part of the family equipment, and the females congregate at the close of the day, bringing their corn stones and corn, and the work is done to the accompaniment of long crooning noises and incantations.

The Patriarch Abraham really began the history of corn milling as we know it, when he told Sarah to prepare "fine meal" for the angels, and in several parts of the "Great Book" we read of "flour," "fine flour," and the "finest wheat flour." It would be interesting to compare the superlative of those by-gone times with a sample of the "Patent" flour of to-day.

For 4,000 years the staff of life was prepared by such rude means as a stone pestle and mortar, a few grains being dropped in the saucer-like depressed surface of one stone and pounded with another of a much similar kind and inverted when in use.

For 2,000 years more the stones were utilised in an improved form, and with a few accessories in the way of sifting and dividing the flour from the bran.

It is only during the last 30 years that pains have been taken to have the wheat thoroughly cleaned before grinding.

It is only during the same period that flour mills have become automatic. Formerly everything had to be done by hand, even to the lifting of grain into the hopper sack by sack.

Another way of expressing the history of milling is by periods.

The period of the hand-stone or quern.

The period of the slave and cattle-driven mill.

The period of the single water-wheel (Greek).

The period of the water-wheel geared to several stones (Roman).

The period of windmills.

The period of steam-driven mills.

The period of cleaning the wheat and dressing the product through silk.

The period of the perfecting of the millstone dress and extension of the dressing process.

The period of the introduction of rolls and a multitude of intermediate processes before the flour was finished.

The perfection of to-day, wherein we wash and clean all wheat, free it from all accompanying impurity, so that when it comes to be operated upon for itself it is entirely by itself. Further than that the miller now understands the chemical and physical construction of grain, and when it has been broken up by the rolls he uses machines for separating those particles which go to make "patents" from those whose constituents are only fit for ordinary flours, and again, that which is qualified for neither is classed "low grade." In addition to these there are intermediate grades. A baker or other customer has only to express his desires and the British miller can so arrange his machines that just that quantity and quality comes into the flour sack.

English milling methods are more complete than those of any other country. It is characteristic of our national build that we never move in any direction until forced to do so by outside circumstances. In a roundabout way corn mills were among the first to employ steam, not as a direct motive power, but to pump water into the mill-dam in order to get a full supply for driving the water-wheels. It was some years before even John Smeaton, the eminent engineer of the latter half of the 18th century, could be made to understand the possibilities of direct driving by steam power. This was, no doubt, on account of the unsteady running of the first make of engines, for stone milling, as well as the roller system of to-day, required just so many revolutions per minute, and those in regular fashion. However, we find that the first steam mill belongs to London, and was started in 1784. Concerning this period, there appears to have been an excellent understanding between millers and bakers, for it is on record that when the assize of bread was in force in the metropolis the millers and bakers managed their transactions in such a way as to assure both of a substantial profit. An instance of this may here be given. The miller would sell flour to the baker at 90s. per sack, but, in order to raise the assize, the price would be invoiced at 100s. This meant the fixing of bread prices considerably higher as a minimum, and when the baker received 10s. discount in addition things were, no doubt, put on a profitable basis. It is quite on the cards that the miller would also be satisfied, as it does not appear he was at this time under any kind of restraint regarding the price he might ask for the flour. Up to the time under notice, there is no mention of wheat undergoing any cleaning process before being ground. It went through the stones " with all its imperfections thick upon it," and we may imagine how it would compare with what is expected at the present date. From now (about 1800) onwards commenced a period of improvement, which is, perhaps, not yet exhausted. It evidently occurred to some one that it might be worth while to get out a little of the refuse with which wheat was contaminated, and we owe to that idea the introduction of the barrel screen—a revolving reel, set at an angle, and covered with wire. Later on, a fan was attached to it to draw away such light matter as short straws, &c. This, again, led to dressing the meal through a much similar device—hence we get flour in contradistinction to what was undoubtedly " meal " before, meal coarser than the ordinary wholemeal of to-day.

For nearly 60 years after the introduction of steam, corn milling—or, as we now prefer to call it, flour milling— again stood still with the exception that fine silk was gradually substituted for wire in the dressing process. English millers date the final revolution in flour milling from the year 1881, when, owing to the vastly superior products that were being poured into the country from

Hungary and America, they were all but driven out of the market.

In that year a great exhibition of roller mills and milling machinery generally was held in London, to which they flocked in large numbers, and, as a consequence of what they saw in full operation, were soon busy remodelling their mills in order to be in line with the foreigner. Most of us are aware of the then superior quality of American flour, and it says much in favour of the pluck that is in us as a nation that in a few years—years, alas, disastrous to a great many members of the craft—that pluck had its reward, inasmuch that I venture to state no other nation would have been enabled to emerge from the conflict at all under like conditions of commercial exchange, but it must be added that when a move is unavoidable there is no nation that rouses itself so thoroughly. This was the case with our millers. The country was inundated with flour from abroad—flour superior to anything that could be made at home, and they were threatened with positive extinction.

In the very earliest times the stones were of all sizes, from ordinary dish plates upwards, and weighed but a few pounds in the original pestle and mortar style. At a later date, however, and when power was invented, they assumed more or less the features which are even now evident in small mills dotted about the country.

In any region of the world the presence of wheat or other cereals, as objects of methodical culture, is a clear proof that man is no longer there a savage. The grains of wheat discovered in the ancient Egyptian sepulchres are no less convincing than the temples and the obelisks, that on the banks of the Nile the people were far in advance of the wandering Arab ; the remains of wheat found in the lake dwellings of Switzerland declare their inhabitants to have long emerged from the state of barbarism. It is to man, or in other words to civilisation, moreover, that the

diffusion of the cereals over the whole world is attributable.
The seeds of other plants migrate accidentally, but the corn
grasses exist only where man has planted them. The
distribution of human food over the earth comes not only
of "natural causes"; there are moral causes as well.

The Hebrews received their corn plants and the arts of
agriculture from a long anterior race. From whom, in
particular, it is impossible to say. Man must have made
the cereals tributary to his wants at a very early period ;
but as to the time of the original sowing, reaping, harvest-
ing, threshing, and grinding, history leaves us totally
uninformed.

The goddess Ceres was so named because she bore all
sorts of fruits. Ceres is called the Law-giver, because,
before men had the use of corn, they lived in the woods
without law or government; but after corn was found out
they divided and tilled the land, and this was the origin of
government and laws. The historian states that the coming
of corn from Egypt to Athens was the signal of great
rejoicings in a time of . famine, and the rites and festivities
were established to celebrate this important event. The
temple of Ceres at Athens, into which only women might
enter, was held most sacred, and the mysteries and sacrifices
which accompanied these celebrations were performed with
much pomp and ceremony. So closely allied to human
subsistence is the growth of corn that it is an uncon-
troverted fact that wherever man can live wheat will grow,
whether in the arctic or tropical regions ; but in any
climate, however congenial, it was never found growing
wild ; it was always the attendant on and never the fore-
runner of civilisation.

Such is the botanical and chemical composition of wheat
that it is not suitable food for man, with his peculiar
physiological structure, in its natural state, either unbroken
or uncooked ; hence it may be fairly inferred that wheat
culture and wheat grinding date back together to those

misty regions of antiquity already mentioned, and all is
obscure and uncertain.

Undoubtedly the earliest efforts of inventive genius must
have been in connection with the milling industry; the
germs of mechanical skill and ingenuity were here propa-
gated in devising means of preparing food. How simple
and rudimentary those efforts were we have the most conclusive proofs.   As the principle of evolution has been
applied in these days to all organic life, we can trace the
same process at work in the history of mechanism as
applied to milling; but we start with this important advantage, we know something fairly conclusive about the "first
form " or origin, and which we take to be that of one stone
pounding or rubbing the grain against another stone, and
then most probably one stone—the bottom one—was
hollowed out, forming a mortar, while the upper stone was
made to form a sort of pestle.   This simple method was in
use for many ages, and was practised by the Romans as
lately as A.D. 79, so that the evolution was very slow.

The ancient history of flour milling, or the making of
edible food from cereals, we can, therefore, venture to
believe, is a most instructive and interesting subject, not
merely to those directly engaged in the milling trade, but
to the public at large, who one and all partake of their
daily bread, either with luxurious accompaniments on the
part of the wealthy, or as the one single course in the daily
menu of the poor and humble.

On the hieroglyphics of that most anciently civilised
country, Egypt, we find representations of the art of
milling grain and sifting the fine meal from the husks of
bran by hand sieves, which must have been a decided
advance on using the crude whole meal.   In their hasty
exit from Egypt, the Israelites probably carried portable
mills with them, for when other provisions failed they had
manna, which is said to have been like coriander seed,
which had to be ground.   One of the earliest laws of

Moses, indeed, recognised the grinding mill as a requisite to every family life. This law is to the effect that " no man should take the nether or lower millstone to pledge, for he taketh a man's life to pledge." In these early days, as in later ones, women undertook to exploit all the latest inventions, and so ousted the sterner sex out of its privilege of making an honest crust, and constituted herself the motive power, as we find several references in the Scriptures to women grinding at the mill. Indeed, to this day in Syria, in Palestine, and in other places, may be seen women grinding at the mill, and there is no spectacle more traditionally poetic and more Biblically interesting than to see one opposite the other turning together their little mill to the cadence of a song. And so right on for several thousands of years, at least, the art of milling in the remoter portions of the country has seen no change ; but it has been otherwise with Western milling, although no doubt these first implements came to us from the East, as we have never heard of anything of the kind having been found amongst the aborigines of Australia to grind the substances used by them as food.

Necessity is often stated to be the mother of invention, but who will doubt that laziness also provokes ingenuity ? The man who got tired of lifting the sacks of wheat up over the millstone hopper hit upon the plan of shooting the wheat down on the floor, and letting it run down into a hole, out of which he made cups or small buckets, attached to an endless band travelling round iron wheels, to lift the wheat and elevate it over the millstones to be ground. Brains gradually thrust muscular action out of the field of the industries, cheapened the cost of production, and marked the progress. It is a fact that the broad principles of flour milling have been practically the same from the third to the nineteenth century, viz.:—The upper stone of the pair was made to rotate over the nether or lower one ; the wheat to be ground was fed in through a hole in the centre of the running stone, when, by reason of the dress or cut, put into the grinding surfaces of the stone, assisted by centrifugal force, the wheat became pulverised on its way to the skirt, where it was collected and sacked off. In later days mechanical sifters and dressers, consisting of woven wire-covered cylinders made to revolve, were utilised for separating the husk of the grain known as bran. Minor points of advantage were applied in some cases, but it is within the memory of adult millers that the above simple method was almost invariably practised.

CHAPTER II.

WHEAT-CLEANING MACHINES.

There is one paramount consideration running through the art of flour manufacture, and that is, Will the flour make good bread? Good bread cannot be produced at random, and the first step towards the desideratum is good flour: flour which will, when supplied with a certain amount of moisture containing gaseous substances, cause the elastic walls of the gluten cells to be distended and the dough to rise, and at the same time become light and porous in texture, pleasant to the eye and palate, and easy of digestion.

To accomplish this there must, throughout each and every operation which wheat undergoes, be a system. There must be some distinct aim in every part of the mill, some goal to be reached at every stage, in order that the whole shall fulfil the desire of those whose inception marks the starting point of the golden berry, and whose eye follows its course through the maze of operations to which it is subjected.

Milling students will do well to disabuse their minds of the idea that there is but one system or arrangement of machines, or that any particular one of the many in existence is right in all its details, because it is idle to pretend that the skill of any one man has exhausted all the possibilities of flow and arrangement of machinery. In fact, a lifetime is too short to grasp all the variations to which roller mill intricacies will lend themselves. That being so, the possibilities of success and error are about equal ; and throughout every day it is the study of men how to avoid the latter by rigidly adhering to the *principle* of milling, and it is to the instillation of this principle that the contents of this book are devoted.

It is not proposed to proceed in the usual fashion and give here a catalogue of wheats and their chief characteristics, as, considering these pages are intended to catch the young miller in embryo, it will be best he should learn in something like sequence the duties which fall to him in his passage through the mill. It is taken for granted that as soon as he has learnt to handle a brush he will be turned loose into the screen house, and here it is best to follow him and give him the ideas he searches for when he views the various appliances at work extracting the refuse and impurities to be found in wheats from almost every part of the world. The screen house is an excellent place to learn some of the essentials of modern milling. It is the very centre upon which the success of the mill depends. Wheat cleaning is almost a science in itself, as there are machines for every form of admixture and which operate upon the various principles of weight, size, and shape, and no aspiring young miller ought to consider it degrading to have such a useful experience placed within his reach as this department affords. To state it roughly, wheat is mixed with various kinds of seeds, oats, maize, barley, straw, rye, cockle, sand, mud balls and stones, and it is the duty of the screens to get them out and leave the wheat perfectly clean. To accomplish this it is hardly necessary to state that several operations are required which demand strict attention on behalf of those in charge. If we begin with what is called the warehouse separator, we shall find that its primary duty is to get rid of a number of impurities larger than wheat, and this it accomplishes by the aid of air currents and sieves. As the wheat is being fed on the machine it is subjected to a powerful exhaust. There are usually two sieves, and the top one has a very

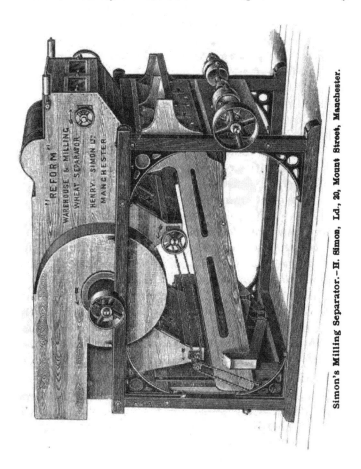

Simon's Milling Separator.—H. Simon, Ld., 20, Mount Street, Manchester.

coarse mesh so as to allow the wheat to fall through; the larger impurities only being tailed over. After passing through the top sieve the wheat is delivered to the bottom one, and on this the meshes are much finer, so as to allow only such impurities as small seeds, sand and dust to pass

through, and in this instance it is the wheat which is tailed
over, and again subjected to a strong current of air as it
leaves the machine. The amount of suction or exhaust
required can be applied by adjusting the air-valves to suit,
and what is lifted out can be examined as to whether good
wheat is being carried away as screenings. The fan will make,
perhaps, 600 revolutions per minute and will require to be
well oiled. The eccentric or shaking device which gives motion
to the sieves must run perfectly noiseless, or else there will
soon be a smash caused by undue wear. Should there be
the slightest "knock" it means that there is too much
play in the box, and the incessant concussion will very soon
wear the eccentric out of its true shape, and then the
wheat corns will dance about and jump off the sieve and
cause a lot of inconvenience. The top sieve will be known
as No. 14 or 15, and the bottom one 8 or 9, preferably
made of zinc, and the motion will be supplied laterally,
that is, from the side; as experience has pronounced this
to be the most successful in every day operation. When
set up true in every particular this machine takes very
little power to drive, and, with ordinary care, lasts a long
time. It should be noted that the feed spout delivers the
wheat the full width of the sieve, so that the separations
are effectual and superior. Good results will follow when
applied to a thin even stream. A first-class machine of
this type is the Invincible Separator.

Another machine of this class contains but one sieve,
and is primarily used on foreign wheat just arriving. It is
a sort of preliminary separator to treat the wheat previous
to being stored, and is a very effectual appliance. Yet
another machine of this class calls also for special attention. It has a small preliminary cross sieve and four full-
sized ones, and usually serves the double purpose of
separator and grader. It has all the attributes of the
others, and in addition lends itself to the work of grading
for the next process. The small cross sieve is intended to pass

everything larger than wheat, such as sticks and straws, maize, string, &c., and the next two are engaged in sliding away such irrevalent material as oats and long barley corns,

Invincible Receiving Separator.—Invincible Grain Cleaner Company, 38, Seething Lane, London, E.C.

and is usually covered with about No. 12 zinc.   Next comes No. 10 or 10½, which divides the wheat into the two sections of large and small sorts, the former going to the barley and oat cylinders and the latter to the cockle and seed cylinders.

There is yet one more sieve to account for, and this is found
doing duty as a sort of sand screen after the division of the
wheat above alluded to, that is, that while the larger sized
berries are sent straight to barley cylinders, the smaller are
yet sieved again to eliminate the sand and small grass seeds
which have escaped the first current of air, and, having
done this, they are again exhausted as they finally leave the
machine.   This machine has a very wide reputation among
small millers on account of the diversity of its operations,
fulfilling, as it were, several functions and thereby up to a
limited capacity doing a double duty; in fact, a few years
ago it was not at all uncommon to find it the sole representative of the wheat cleaning plant, especially where much
clean English was milled, because it must be borne in mind
that native sorts as a whole are about the cleanest samples
in existence.   Neither do machines of this class need a
great deal of attention.   They are not adjustable except
to aspiration, and the great thing is to keep the sieves clear
of accumulation and the working parts clean. The screening
box should be constantly examined to see that no good
wheat is being wasted in that way, and, on the other hand,
it is advisable that as much as possible short of overdoing
it may be accomplished.   This refers to every machine in
the screen house.   Keep the feed up to quantity fairly well,
and get all the refuse away ; do not depend upon the next
or any following machine to make up for lost oppor-
tunities ; look out for loose screws and nuts on the machines
themselves, because the constant vibration is all in favour
of such things ; oil well and keep clean, and satisfaction is
sure to accrue.

If the wheat be not graded as just stated, but is treated
with a separate appliance, then nothing can be better than
an ordinary sieve having a rotary motion, and fitted with
covers to correspond with the sizes required and the number
of them.   This can be carried to almost any degree, as an
ordinary blend of wheat will sift out into many sizes and

the impurities will do the same. For our present purpose, however, we will divide it here into only two, viz., large wheat over 10 or 10½ zinc to oat and barley cylinders and throughs to cockle cylinders. The object in view is to get

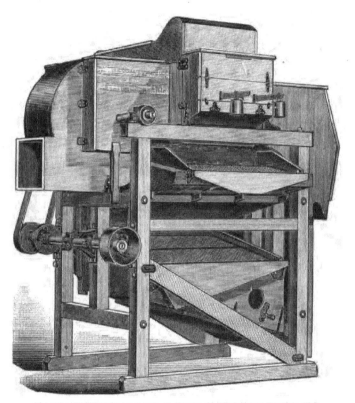

Eureka Milling Separator.—S. Howes, 64, Mark Lane, London, E.C.

a division of the good from the bad by size and shape. An examination of the interior of the former cylinders will show that the drilled holes are intended to pick up the wheat and hold it until its centre of gravity is disturbed, and it gets pitched into the internal trough. As a general

rule oats and barley are longer in the berry than wheat,
and cannot accommodate themselves to the size of the
cavities, and it is this fact which the engineers have taken
advantage of in construction.　It must be quite plain that
if they cannot follow the wheat they must find an outlet
by themselves elsewhere, and this is just what happens.
On the other hand the small wheat in the other cylinders
takes the place of the oats and barley in the large wheat.
Here the holes are comparatively small, too small for wheat
to stay there, so that it is left alone, and the seeds and
other small refuse make their exit by the internal trough.
That is really the principle of all cylinders, and a clear
understanding what their particular work is ought very

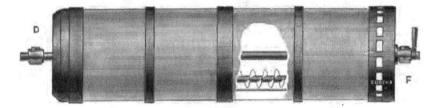

Eureka Cockle or Barley Cylinder.

soon to make any one conversant with the mode of
regulating the angle at which the movable trough shall be
tilted, so as to get all possible refuse away and leave all wheat
behind. The detached " pickers " inside will want renewing
periodically, as will also the inside binding rings of the
cylinders themselves, but with attention they will last for
years as a whole. There is, of course, a certain amount
of wear and tear, and the perforations are in time worn
too large by the action of the continual tumble of the
wheat upon their surfaces. The revolutions are from 10 to
20 per minute. They are from 14 to 40 ins. in diameter,
and from 4 ft. to 12 ft. in length. There are special
perforations made for the elimination of rye upon the same

principle, but enough has been written to show the observant man the path to increase of practical knowledge by manipulating the parts already put before him. Suffice it to say that some of the latter class are adjustable, and that by previous careful grading it is possible to obtain a very good separation of this objectionable admixture, which, if not removed, gives the flour a more or less leaden appearance and lowers its strength according to the amount allowed to pass through the milling machines proper. It may be here noted that the bulk of loose impurity is supposed to be eliminated by the time wheat has passed these machines, the rest of what is called the dry process of cleaning being devoted to detaching adhering increment, either natural or accumulated, and the more perfect the earlier operations the more easy this is of accomplishment to a degree determinable only by knowledge obtained among the machines themselves, backed by a growing experience of the wheats operated upon.

Scourers come next in order, and are of various designs. Some are adjustable—that is, the internal mechanism can be moved to do more or less duty accordingly as the wheat demands it. The only idea in this department is to free the wheat corns from adhering impurities and to break up loose dirt lumps, and whether any make of scourer accomplishes this or not is the only true test of its efficiency. A fan is always attached to carry away liberated refuse, and the deposit in the screen box is a guide to regulate the current by, while the state of the wheat after the operation testifies to the severity or otherwise of the process undergone. Whether the type be vertical or horizontal, attention must be paid to it so as not to allow the wheat to be subjected to a preliminary first break, that is to say the berries must be intact as they proceed to the brush machine unless indeed the state of the wheat demands such drastic scouring by being smut-ended or otherwise befouled. It will need close observation, working

as it does upon many varieties.   Its action is not governed
by the laws of any previous machine, and serious damage
can be wrought through inattention.   The internal beaters
attain a speed of 500 or 600 revolutions per minute, and
the effect can at any time become aggravated if they are
allowed to be set up too close or the amount of feed be
greatly in excess of capacity.   Sometimes this may be.
noticed in samples of English wheat coming to the mill.
Numbers of corns will be found split open owing to the
too close setting of the drum in threshing machines, and it
always discounts the intrinsic value of the whole because
there is the likelihood of their being extracted during the
cleaning process and swelling the amount of screenings.

CHAPTER III.

WHEAT-CLEANING MACHINES.

Brushes are of various kinds and an indispensable machine in every mill. Whether washing plants are employed or not, this machine finds an honoured place in finishing up the preliminary process which wheat undergoes before it is considered to be in a fit state to be operated upon for the purpose of extracting the flour. The several makers claim all that is required for their machines of this name, and it is no set purpose of mine to put any single stamp in the forefront. Each have undoubted good points, and all I remark is that the results must be such as to show, on examination, that the wheat is what it is expected to be after passing through them. Hair, or beard, as is termed the fuzzy portion which grows on and adheres to the small end of wheat, ought to disappear during the operation, and bits of thin transparent tissue brushed from the wheat's surface coat and exhausted away, and the berries should issue forth almost glistening in appearance as a result of contact with the revolving fibre encased in the framework. If it be not so, then it is evident that the brush surfaces have worn away and require to be set up closer to the inner casing, or if they have acquired an uneven surface they need to be trimmed level again, and this applies to both solid, segment, and cone machines.

Magnets are largely employed in English mills to extract pieces of wire, nails, and sundry other iron and steel

particles which unaccountably find their way into the
wheat. The first-named is most likely to emanate from
the mechanical binders employed in the harvest field,

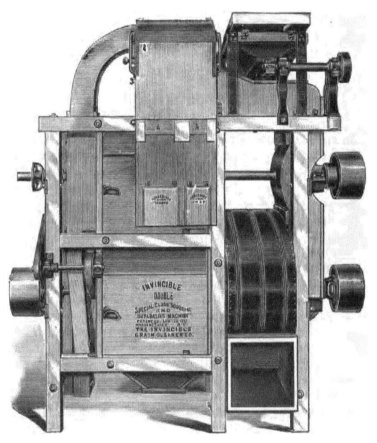

Invincible Scourer.

though of later period it has nearly been discontinued and
string used in its place. Magnets consist of either a highly
sensitised plate over which the wheat is made to slide, or

of horseshoe shape inserted at the top of a spout, and which draw and hold their affinities until the latter are removed by hand, or automatically, which should be done regularly to avoid accumulation and consequent dislodgment of some of them. Where one set only are used, they will in most cases be found somewhere in front of the scourer, so as to avoid the risk of that machine causing sparks by violent contact of the beaters with them. Many millers, however, employ them also just over the first break

Eureka Brush Machine.

to make sure that nothing of that nature is going to mutilate the corrugations on the rolls. This is a highly sensible plan, and will fully repay the slight outlay involved.

I am afraid, however, that in a great many cases the screening of wheat is only carried out on general principles which more than likely are out of touch with the high ideal we are supposed to be aiming at. It is of vital importance that every particle other than good sound wheat should be removed, and this in reality is the first step towards producing good flour, free from cast or speck. We have even yet something to learn in wheat cleaning, and really it is here that the first departure ought to have been made; but instead of that millers began to use rolls on imperfectly cleaned wheat, and are only now fully alive to the fact.

Automatic Grain Weigher.—W. and T. Avery, Birmingham.

Automatic weighing machines are a feature in mill warehouses, and during the last few years a good deal of attention has been paid to them by inventors to make them as reliable as science will permit. The result is, on the whole, very satisfactory. They have their properly-appointed places, which appear to be—first, to check off any parcel as it arrives and before anything is abstracted

in the way of impurities ; and afterwards, when the wheat
has undergone a thorough cleaning, the process is repeated
to ascertain what loss has been entailed in its travels
through the screen-house. This gives a somewhat
accurate idea of the gross value of any one sort of wheat
so treated. The nett value is, of course, calculated upon
what percentage of flour is obtained. In mechanism they
differ slightly, but the principle is identical in all makes.
They require being kept clean in all parts, so that the
delicate balance may be maintained. During the period
occupied in filling the receiving hopper the discharge must
be seen to be shut off absolutely, and during discharge
the stop brush must act with perfection in keeping back
supply. The numbers on the discs being copied off
previous to starting and deducted from the aggregate
gives the quantity of discharge per hour or per day as
intended. The slower machines with large hoppers are
mostly favoured, as having less wear and tear than those
which seem as it were to be in a constant hurry. I consider the future will witness a more general usage of them,
and a study of their component parts will benefit all who
give some attention to their ingenious construction.

I have not before touched upon dust collecting devices
for wheat cleaners, as I preferred to shortly review them
here. There is only one principle involved, and that is
this : After the fan has done its duty in eliminating the
lighter particles from the wheat under treatment by
sucking them away, it remains for these particles to be
again dealt with and separated from the current of air
which the fan, by creating this partial vacuum, was
responsible for. To accomplish this various devices are
called into requisition, and yclept dust collectors, it being
understood to mean that by their action, or in their construction, or both together, they so manage that the dust
and chaff, being the heavier, shall be deposited, and the
air allowed to escape free and comparatively clear, and to

the extent that this is done by any given apparatus of this
stamp are they judged by those who use them.   It must
be understood that it is a difficult matter to compass without curtailing the effectiveness of the fan.   There is always
a suspicion in these self-contained machines of the area

Dell's Victor Brush.—W. R. Dell and Son, 26, Mark Lane, London, E.C.

being insufficient for the free outlet of the air which has
just done duty, and is anxious to escape from the pressure
behind it.   Once let it be proved that this back pressure

does not exist, or not to a degree interfering in any way with a machine's efficiency, the makers ought to reap the reward of their ingenuity. All claim this immunity from back pressure, but my experience only brings to mind one or two at most which nearly fulfil this condition.

In this part of the mill textile dust collectors have largely given place to machines of iron, zinc, wood, or a combination of them. Flannel and cloth are of too inflammable a nature to pass muster with watchful insurance companies, and are only rarely met with in these days. On account of its natural density and the fineness of its particles, wheat dust is to an extent of an explosive character; hence anything which conduces to the spread of an accidental spark is avoided as far as possible. Textile material of the kind employed will not, it is true, flare up, but will smoulder— when not at work. But let a strong current play upon it and a blaze will not be far away, or at least an approach to one will be near enough for fire purposes. Of the non-textile kind suitable for use in the wheat-cleaning department there are several, including the Cyclone, Tornado, and one or two others, all well approved and at work in all parts of the kingdom. They need no explanation, following for the most part the natural laws of air contraction and expansion and not being burdened with a lot of moving mechanical parts, it needs but the merest tyro to understand that air, just previously compressed, on being liberated expands and attempts to rise, and should there be no opposition it will do this and deposit anything of greater specific gravity than itself into the receptacle provided for this contingency. By blowing the dust-laden air into a specially-constructed appliance of greater area it —the air—loses its momentum and therefore its hold upon the dust particles, which of themselves fall away while the air escapes upwards.

To review what has been written in a concise form, the separator will sometimes be found striving to do its best

under unfavourable surroundings. Perhaps in small mills
it is the sole representative of the cleaning plant, and, in
addition, may be screwed down in an out-of-the-way corner
—away from light and ventilation, as if it were a sort of
necessary evil following in the train of a new roller plant.
It is nobody's duty to attend to it, and it is more the rule
than is acknowledged to let it severely alone. When this
is so it is bound to tell on the flour, which will not hold

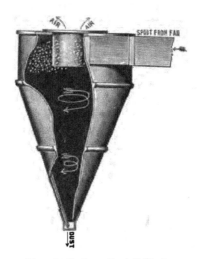

Simon's Cyclone Dust Collector.

its place when such incipient carelessness is allowed at the
very commencement of its travel. Skilful handling of the
rolls or purifiers will not atone for omission in attending
to, or undue shortening of, the wheat cleaning plant; and
it may here be stated that this blot upon millers is
answerable for a lot of under-class flour put upon the
market which does not fetch the price the mixture would
otherwise bring.

There is no finality in wheat cleaning. So long as the
berry be whole it is continually shedding the fine powder

from the husk, and every time it is stirred the process is accentuated. The keeping up of a continuous frictional operation during the whole period of storage, accompanied by a current of air, is about the only thing at present known as feasible to obtain absolutely clean wheat.

There is no standing still as it were, after being harvested and stacked, this process of outward decay seems to at once begin, and may be caused to some extent by the slight shrinking of the berry as it hardens or matures. Whatever the cause, the effect of bran powder is always deplored, and to get as much as possible away by trituration before the actual process of flour-making begins is rightly considered to be of the first importance whenever and wherever a high brand of wheat product is intended for successful competition. Cockle and barley cylinders are open to the same complaint, and it is not seldom that they are seen working very much under their maximum of efficiency.

Unfortunately, perhaps, their chief operation is upon the most expensive sample of wheat, and to see a few grains tailing with the wheat is mostly considered a grave offence. We are all of us of the gender economic in this respect, but when the subject is well considered, one grain of small wheat among the cockle screenings is twenty times better than a grain of cockle left in the wheat, and with the most ordinary attention nothing serious need be apprehended at this point, as these machines will stay as they are set, and the feed to them is not so varied as to need alteration every few minutes. Barley cylinders are in the same category, and a slight excess is to be preferred to the plan of overcarefulness.

Although not so mischievous in their effect on the flour as the black impurity from the cockle cylinders, it is better that the finished product should not be weakened by the kernels of oats and barley when the means of prevention are ready at hand. Taken in its entirety the horizontal type of scourer is perhaps more favoured of late years, but, if fed too heavily, the power consumed becomes excessive, causing slipping belts, chokes, and too much wear and tear. These machines should be adjustable and have plenty of wind power attached, because in those of the vertical class a good suctional action holds the grain in partial suspension and thus gives the falling wheat a better chance of being acted upon by one or more of the rapidly revolving beaters than would be the case with a quieter gravitation, consequent upon the using cf just sufficient wind to carry away the dust and chaff scourings. With regard to quarter twist drives ordinary leather does not seem to meet requirements; ráw hide or untanned is better, being more supple and consequently more lasting, and that with less intermediate trouble. None of the machines enumerated thus far will give the best results if they are placed in cramped out of the way places. They need plenty of free open space to allow the fan to draw its supply of air, and should there be any difficulty about this or a throttled outlet, evidence will be found in the deposit of scourings beneath. The space between the perforated jacket and the casing is also more than likely to become clogged with dust during damp or foggy weather, or when cleaning moist wheat, and frequent attention will be needed to obviate or avoid any such curtailment of efficiency. All the foregoing remarks relating to aspiration are also pertinent on the generally remaining dry process of wheat cleaning, and that is the brush. The brushes must be kept well exposed to within a short distance of the perforated covering, and the fibre trimmed occasionally, and there is sometimes a tendency towards accumulation in the heart of the machine which a slight tap now and again will remove. Small millers do not, as a rule, keep a man on the screens continually, but experience suggests that it would pay far better than keeping flour on hand, because there is no such man on duty in that department.

CHAPTER IV.

Wheat Washing and Conditioning.

It is not long ago since elaborate screening was governed by space and power, and not by necessity, and I predict the same reaction in favour of the operation of cleaning by the aid of water.

I have held for some years an opinion, the result of experiments, that water judiciously applied is the most potent instrument in profit-making that millers can use. In the event of the effects not being satisfactory the fault has been, not in the principle, but in the mistaken application of the details of the process.

We are about to enquire very strictly into the characteristics of various wheats ; to watch their behaviour ; to note advantages and to classify results so far as this can be done.

In order to make my meaning as clear as possible we will divide certain wheats into cerain classes and see how they should be treated.

There are, firstly, some wheats which require tempering besides washing. Broadly, these are Russians, Persians, Indians and Egyptians. I do not say that the above rule is an invariable one, because the mixture of others will have some say in the matter. The kind of flour trade cultivated also must have influence ; and, again, the idea that conditioning is not an absolute necessity in any case. There are wheats also which require washing for washing's

sake. They are dirty. There are wheats which are
damped or washed simply to toughen the bran. Finally,
there are wheats which are already soft, but may be dirty,
such as soft Russians, Plates, Winter, and, perhaps, smut-
ended English. The working principle of all wheat washers
is the same, and they only differ in the way they each fulfil
this principle. The balancing agency is water where the
wheat contains lumps of dirt or small stones. A current
of water is used sufficiently strong to influence the wheat,
and not sufficiently strong to influence the stones.
Stones and mud balls may be, and mostly are, approximate
in size to the wheat itself, but they are heavier, and so
engineers utilise this difference in gravity to effect a
separation.

Resuming, we may take it for granted that the washer
proper is followed by some kind of whizzer, and I will sum
up the character of a whizzer by saying that it is a
machine which combines in itself the double action of
centrifugal force and gravity.

Among the many excellent washers and whizzers is that
of Rowlandson's, which, although it takes up very little
room, has a large capacity, and is at work in all parts of
the Kingdom. A reference to the illustration and the
corresponding lettering here inserted will be found useful
in understanding its chief parts.

Mr. Rowlandson claims that this machine, with a stoner
attached, will wash, clean, and dry wheat without breakage
or abrasion of the bran coat. Certainly it does its work
quickly, and this must be classed as a decided recommendation.

I have mentioned a certain class of wheats in the first
division—namely, Russians (Hard), Indians, &c., where
water is employed for cleaning, mellowing the berry, and
toughening the bran, and generally rendering the wheat
more amenable to the action of the various machines which
together form a modern roller mill. These hard wheats

require water to enable them to mill better, and some go so
far as to say that the flour produced from them is rendered
what is usually termed stronger by such admixture.   They
say they can take flour made from dry Indian wheat and
apply the ordinary test for gluten, and get a small quantity

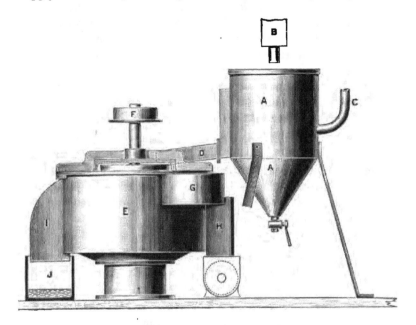

Rowlandson's Whizzer.

A—Stoner (any type may be used). B—Wheat-Feed into Stoner. C—Water
Supply into Stoner. D—Wheat and Water from Stoner into Washer. E—
Washer and Whizzer combined. F—Pulley for driving Whizzer, runs against the
sun. Spindle any length. G—Dried Wheat Delivery Spout into Worm or
Hopper. H—Grain Spout to Worm Hopper. I—Wind and Water Outlet. J—
Discharged Water Outlet.

of very inelastic gluten; if the same wheat be damped
before being made into flour, or if the flour is thoroughly
damped before the gluten test be applied, gluten greater in
quantity and of superior quality is obtained.   This I had
rather not dilate upon.

c

One of the most important factors in connection with wheat drying is often unrecognised or ignored, yet it has a most important bearing on the treatment of soft wheats, and explains many of the hitherto unexplained discrepancies in the working of wheat-washing plants apparently exactly the same. The fact to which I call special attention is this—that the action of the heat when applied to wheat differs widely in its effects in proportion to the amount of moisture the wheat contains. Thus, if very wet wheat be subjected in a modern dryer to air heated to 200 degs. F., this wheat becomes very hot, but if dry wheat be put through the same machine with air at the same temperature it becomes only moderately warm.

To illustrate this we will say that Russian wheat, washed and left very wet, is put into a dryer and subjected to the action of air heated to 200 degs.; this temperature produces so much steam from the moisture on the wheat that in effect the wheat is treated in a steam bath, and instead of being dried is actually softened.

But if the wheat be almost free from moisture, no steam is generated, and the heated air is used for its legitimate purpose, and the wheat dried and the temperature kept 30 to 40 degs. lower than with the wet wheat.

These hard wheats are again benefited in respect to what we generally call flavour in flour. Indian wheat should not, as a rule, be thoroughly dried immediately after washing, but should be allowed to stand in sacks or bins until a very slight fermentation takes place. If this be carefully done all trace of earthy flavour will be lost and the flour will be undistinguishable, so far as flavour is concerned, from that made from American Winter wheats. The same process varied to suit circumstances is suitable for use on all other hard wheats.

All hard wheats are very difficult to bring into milling condition without an interval between some of the opera-

Wheat Dryer and Conditioner.—T. Robinson and Son, Ld., Rochdale

c 2

tions of cleaning, washing, and drying. Perhaps the best method, the process which by its slowness of operation is best suited to these wheats, is the old-fashioned tiled kiln. The writer has seen many thousands of sacks of Egyptian wheat brought into excellent milling condition by its use, but this is both too expensive and too slow for use in modern automatic mills, and perhaps there is no better practical method than washing, draining, and whizzing them very slightly, letting them stand in sacks for some hours, and then again putting them through the ordinary course of washing, cleaning, drying, and cooling. Of course this costs more to do, but the wheats that require it are usually fairly cheap in the market and will stand a little extra expense in manipulation, especially when the results are so much improved.

Hard Taganrog, if it has to be used, may be treated in this way, but to mill this wheat, or other wheats whose texture is horny, most advantageously, is better done by making a few special alterations in the flow of the mill to suit their character than by endeavouring to work them by severe pressure of rolls. Heavy pressure on wheats of this kind means a large increase of power without a corresponding increase of flour production, and should be avoided if possible.

The second division may include the softer Indians, Russians, hard Americans, such as Spring, Duluth, Manitoba, Kansas Winter, Californians, Chilians, and Walla, and in some seasons hard Plates. Most of these wheats are benefited by the use of water and heat; they are rendered cleaner, and the bran is toughened and less liable to trituration in the break rolls, the percentage of broad bran is increased, and the break flour and the later reduction flours are clearer; while the patent flours are free from specks.

Some of these wheats will bear a slight addition of moisture to the endosperm without injury, and may be treated on an ordinary washing plant without much danger of the effects of water being found in the flour sack. The Californians and the Indians which come under this division, such as soft white Bombays, Calcuttas, and Kurrachees, may sometimes be left for a short interval between washing and drying, but it is probably the safest plan to let the dryer immediately follow the whizzer, and thus get rid of any adherent moisture, and let the toughening of the bran be done by the slight steaming action of the dryer. If this be followed by a cooler these wheats can be taken from it and stored in sacks, if need be, in perfect milling condition.

It is probably in the treatment of the wheats included in the third division that the most serious differences of opinion arise amongst millers, and the most serious mistakes are made in treatment.

All millers are now willing to agree that for treatment of Indians and harder wheats water is a necessity and heat not detrimental ; but when one suggests that soft Ghirkas and Plates should be washed it is quite a different matter, and doubts are frequently expressed as to utility and safety. Yet many of these wheats are very dirty ; small hard earth balls, smut balls, and broken smuts are very common impurities contained in them—impurities that no process of dry-cleaning can perfectly eradicate, but with water their removal is simple and easy.

For the treatment of these soft wheats neither of the two previously-mentioned processes are quite suitable, and the first is absolutely fatal to beneficial results when applied to soft wheats.

For the best, or indeed the only real efficient and correct process of washing these wheats, three things are absolutely essential :—First, that no more water shall be used than is essential for washing ; secondly, that the wheat shall be immersed for the smallest possible time, a few seconds only, and that any water left on the wheat

berry should be immediately and effectually thrown off by
the whizzer or whizzers before the wheat enters the dryer.
If these points are all carefully attended to, it is possible
to wash soft wheats without in the smallest degree affecting their condition or impairing the strength of the
resultant flour.

Washing hard wheats and washing soft wheats are two
distinct processes and should not be attempted on the same
lines. In the hard wheat, a process of conditioning is an
advantage, a small addition of moisure is also an advantage, and heat up to a moderate degree, probably higher
than most people think, is not injurious; it is to this
increase of heat that the conditioning effect of most
modern dryers is due ; wheat containing a certain amount
of adherent moisture is subjected to the heat of the dryer,
the effect of that heat is to change that moisture into
steam, or something very like it; this steam penetrates
the bran, fills the bran cells with moisture of a fairly high
temperature and softens and toughens them, and also sets
up a slight sweet fermentative action, which is distinctly
beneficial. In a soft wheat even the smallest addition of
moisure is harmful, the berry does not require conditioning,
being already in good milling condition, and also being
more susceptible to the effect of heat than the thicker-
skinned Indians; hence the importance of the points
mentioned.

Condensed the washing process is this : Complete
immersion to wash the wheat and also to separate loose
impurities, such as stones and hard mud balls, usually a
draining worm set at an angle to enable the water to be
drawn away, an horizontal or vertical whizzer to throw
off surface moisture, a series of hot plates down which
the wheat slowly slides, and which softens the outer integument of the bran tissue, a scourer to remove this tissue by
the action of its beaters and through another series of
plates, this time perforated as a rule, to allow of a strong

current of cold air being drawn through them, and consequently through the descending stream of wheat. A final
brushing operation completes the circuit, and if all are
regulated with the necessary skill the wheat will be ready
for mixing, and can go direct to the first break roll for
immediate conversion into flour.

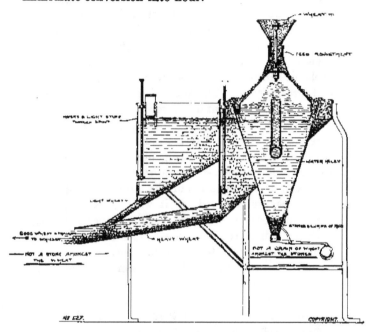

Wheat Washer and Stoner.—E. R. and F. Turner, Ld., Ipswich.

The system of getting wheat into the mill may be briefly
described before we leave this part. In small and country
places it is usually by sacks which are hoisted by chain
tackle to some convenient floor until wanted for mixing,
and then they are shot into a bin or silo as per instructions.
The same operation is gone through if the mill is fortunate ·
enough to have railway facilities where they can be hoisted
from waggons direct. It is quite another matter where

grain is brought within reasonable distance by steamship, and although strictly speaking it is not a technical subject connected with flour making proper, its mechanical operation is necessarily complex to a degree when viewed for the first time, and a few words of explanation in regard thereto will not be out of place. In this instance we will suppose that a boat laden with wheat is alongside the warehouse front, a movable elevator—that is, an elevator that can be raised and lowered to suit the lowering of wheat caused by a lessening bulk, and ebbing tide, or any other probable contingency—is set to work where it can pick up and load its cups. This done, it delivers on to a travelling band running horizontally towards the mill building. By what is known in engineering parlance as a throw-off carriage the wheat is diverted in almost any desired direction, either straight to storing bins or through automatic weigher and warehouse separator previous to being sent to bins, and by the adoption of the latter a lot of the lighter impurities are removed, as it were, first hand, and by removing pieces of straw, sticks, and other accompanying particles, much trouble is saved when the wheat comes to be measured out in the revolving pockets of the mixers, which are usually placed at the bottoms of these receptacles for bulk samples, and the fan on, or attached to, the warehouse separator does very good duty by eliminating the dust and so helping to keep the wheat in good condition.

The travelling band mentioned goes at the rate of 400 to 500 ft. per minute, and the speed is limited to about the latter figure on account of the displacement of the surrounding atmosphere. If a much greater rate were attempted horizontally it would be found that the wheat would not stay on the band, as the pressure of air rushing to fill up the partial vacuum would scatter it, and this principle alone governs the speed at which wheat or other material of like gravity shall be so transported for any considerable distance.

Some millers even study individual wheats. By this I mean that a certain wheat will be damped or washed and whizzed and then allowed to stand for, say, eight hours;

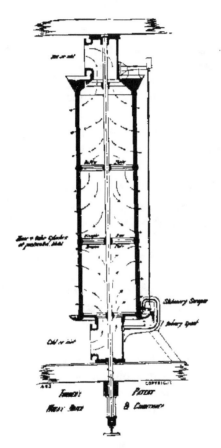

Turner's Wheat Dryer and Conditioner.

others will require 10 or even 12 hours' rest, and there are also wheats which require but two hours. The writer has seen mills where this is practised, and with the highest results. This points to the fact that we are, as English millers, progressing beyond the confines of the ordinary almost every day. There is another side to this, and that is the time, trouble and expense, and many millers consider it both too slow and too costly for modern mills. Many things have been spoken and written on this subject of wheat washing and drying, but, after all, perhaps the general purpose is mostly fulfilled by washing and whizzing in conjunction with each other, and without any interval between. Here is a list of machines required in a first-class merchant mill :—

1. A receiving elevator and accessories, such as spouts, band and connections to deliver all or any wheat direct from quay or dock straight into the mill.

2. A reliable self-registering automatic weigher.

3. Warehouse separators to give the wheat a preliminary cleaning in order that it may stand prolonged storage if required.

4. A distributing indiarubber band to direct the wheat to the silos or bins.

5. Separators or separators and graders to further clean and size the wheat for the cylinders.

6. Cylinders for the extraction of oats, barley, cockle and rye.

7. Scourers to break up hard refuse matter and smut balls, fitted with strong aspirators to carry off all products so loosened by the operation.

8. Washers and stoners.

9. Whizzers to throw off extraneous moisture.

10. Machine of the heater type to dry the coat of the wheat, to draw out excessive moisture in the case of damp wheat, and to desseminate inherent moisture in the case of wheats in an opposite condition.

11. Dryer or conditioner to cool the wheat quickly, to

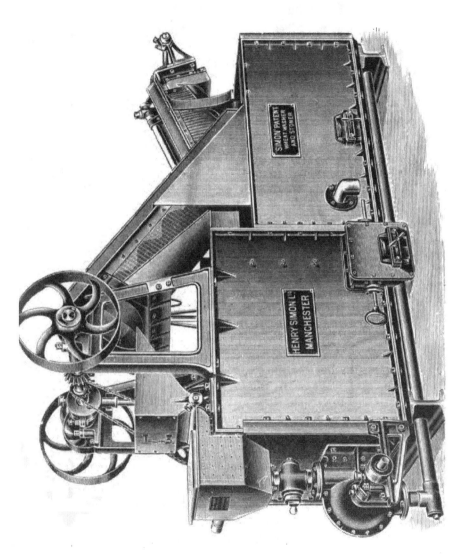

Wheat Washer and Stoner.—Henry Simon, Ld., Manchester.

bring it to a normal temperature before damage has been done by the heat and moisture employed.

12. Brush machines to give a final polish before being sent to bins for mixing purposes.

13. Slow running fan of large diameter, exhausting from all the silos continually.

14. A line of wheat mixers underneath all silos to measure out the required proportion into a receiving worm.

15. Magnets to be placed in spouts where experience points as best calculated to arrest all particles having an affinity therewith.

16. Automatic weigher to weigh the clean wheat as it proceeds to the first break rolls.

17. A grader to grade the wheat into two or more sizes to suit the different corrugations on the initial breaks.

A mill fitted up on these lines would, if wanted, successfully operate even upon the worst sample of wheat which ever emanated from the land of the Pharoahs, if the machines were not overfed. This overfeeding of nearly all flour-milling machinery is an error. There is a point at which all properly-sized and properly-equipped machines will do good work following each other; but the capacity varies with different wheats, and it will never do to keep the same feed with wheats of varying qualifications. Attention to various sorts and mixtures will soon make a man adept in noticing the little differences, and it is this habit of attention that awakens an interest which, if closely followed, is bound to end in the betterment of his own position, as it undoubtedly adds to the intrinsic value of all products upon which it is lavished.

Another way of explaining the action and the purpose of wheat-cleaning machinery would be to say that it exists solely for the following duties, viz. :—

1. To extract everything larger than wheat, such as maize, excrement, sticks, straws, string, &c.

2. To extract everything smaller than wheat, such as sand, seeds, &c.

3. To separate particles of nearly the same size and specific gravity, as barley, oats, and rye.

4. Pieces of wire, iron, or nails.

5. Extraneous matter, such as light dust, chaff, and also weevilled, dead, or immatured grain, and smut balls, all lighter than wheat, which is sound and intended for use.

6. Extraneous matter heavier than wheat, as stones, mud balls, or hard dirt, and the like, so that three principles are involved—namely, separations by size, gravity, and magnetism.

If it is at any time necessary to describe the passing of wheat through a full line of machinery, the following may be of service :—From silos to milling separator for the removal of large impurities, of dust and sand. Graders are next in order and divide the wheat into two or more sizes for the cylinders, and here we make a deviation in the case of wheat about to be washed. From the cylinders we usually send it direct to washer and whizzer, and, whenever practised, to steamer or heater, to cooler or dryer and finally to brushes, and then it is supposed to be ready for the mill proper.

In summing up the question it may be said that all the machines are supposed to be automatically connected, one with another, and, should it be preferred, an automatic weigher is placed just above the first break to register the amount of clean wheat operated upon, and the time expended on the collective machines will not exceed 12 to 16 minutes in a well equipped mill.

If, as just mentioned, the clean wheat be weighed immediately before being reduced by the rolls, as it was weighed before being treated by the warehouse separator, a perfect knowledge of the net loss is always to be obtained. An arrangement for exhausting from clean wheat bins and also

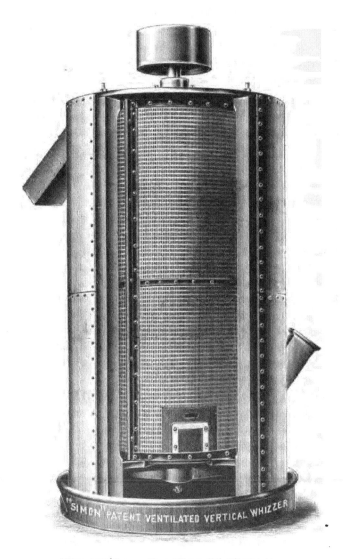

Wheat Whizzer.—Henry Simon, Ld., Manchester.

as the wheat is being graded for first break completes the circuit of wheat preparation as practised to-day.

In almost all wheat mixtures one definite object is kept in view, and that is that the wheats used shall so combine in their various qualities and quantities that the result shall be satisfactory to the customer and profitable to the miller; and to condition wheat so that the whole shall best fulfil the object aimed at. The physical constituents of wheat are presented to us by chemists as being pretty much alike in their general characteristics ; where they differ most is in the gluten department—that is for milling purposes— and in the structure. This is mostly owing to the influence of soil, pedigree, and climate. A miller is supposed to know pretty well how to mix from various sources of supply ; but wheats are so vastly different in so far as regards being hard and soft, clean and dirty, weak and strong, that of late years appliances have been devised whereby the manu-facture of flour from wheats widely divided in their several qualities is made easier of accomplishment. By their combined attributes they so act that all may be brought into line with each other, and blended, so that when they reach the mill proper they stand equal chance of being separated into the various concomitants so familiar to practical men of to-day. Thus, a soft sample is operated upon by conditioners for the sole reason of vapourising and drawing away excessive inherent moisture—that is, taking it as in other respects perfect.

A clean hard wheat is first washed, and then conditioned for the sake of imparting a mellowness it would not other-wise possess.

A dirty sample of either sort is perforce washed for its own sake, and afterwards so treated by heating, scouring, and cooling that a transformation takes place which enables t to be put beside any other variety for milling purposes.

Other sorts used have extraneous matter mixed in them which no other mode of cleaning will free them from, and,

as a necessary consequence, must be sent through the same
operation to either correct their natural infirmities or to
prevent serious damage to their specially valuable properties
—gluten, for instance—and the main principle underlying

· Wheat Whizzer.—T. Robinson and Son, Ld., Rochdale.

the whole process, not allowing for profit, is to so manipulate
the machinery that when they have got through the ordeal
soft wheats shall be more concentrated, hard wheats shall
be more mellow, dirty wheats shall be clean, and all shall

show the proper amount of benefit derived therefrom. Should this not happen to be the case, then the various points which the wheat passes need to be examined, so that inlet and outlet valves on the conditioners are properly set to allow the feed and outlet a steady regular flow; steam to be generated and circulated at a regular temperature, slipping belts (possibly) attended to, feed to be constantly inspected, and level of water kept regular. An intelligent man would at once note the variations mentioned in the question, and a thorough knowledge of the particular machines under his charge would make him alive to little peculiarities not specified, but which are localised to his particular mill. Such little changes are bound to occur, and mere generalising will not meet the case, perhaps as exact as being in personal touch. This, at least, is the usual experience, the only thing being a general cognisance of the operation and the results expected, and this it is makes a man more alive to make points which he knows to be susceptible to these changes his special study. To sum it up, it remains simply a question of speed and feed, and strict attention to both will, all other things being equal, compass the desired end.

Wheats which have just undergone the ordeal of washing and whizzing, as being necessary operations, are, as a rule, in that state which calls for the exercise of high intelligence to put them just in that condition which will be of the greatest benefit. The science of drying consists of a knowledge of the multitudinous varieties of wheats imported, their principal characteristics, such as gluten tests, amount of natural moisture, the primary reason for their undergoing the whole or part of the process, the condition they are wanted to be in for milling purposes, the dangers which are apparent should .the right treatment not be given to them, and the result of such a want of knowledge as shown in the flour or bread. For instance, soft wheats of native or other

growth will be conditioned or dried for the sake of
extracting the moisture—or some of it—so that they
will mill more freely, make better and clearer middlings,
and more of them, and thus lessen the quantity of break

Robinson's Wheat Washer and Whizzer Combined.

or other lower grade flour which results from premature
conversion before the preliminary processes have been
got through. As a rule they are of the weak kinds,
and danger to gluten is not so much taken into account
as when strong, hard samples are being operated upon.
The latter require far more attention, because they are
indispensable in every mixture on account of the gluten

they contain. It may be that the principal reason for having to go through this process is because of the loose dirt originally present, or it may be impurity attached to the grain itself, or, again, if the wheat be clean in the berry, the structure of it may call for its being conditioned. Whatever the cause, it will require handling with care. Too long immersion may penetrate the outer covering, and start something slightly akin to germination, and so spoil all the effect aimed at in de-grading the quality of the gluten. Shortly, the science is this: To bring all kinds of wheat into that state best fitted to be made into flour as we know it to-day and to avoid injury to any vital constituent, and this can only be obtained by a study of the wheats of the world, and how to best treat them according to their various qualifications for flour-making purposes.

Atmospheric conditions also enter somewhat closely into this question of wheat conditioning or drying, as they always did before. No one of the older millers would allow wheat to be turned on a damp, foggy day, as the belief was (and rightly so) that it did more harm than good on account of the contact such operation entailed with the surrounding atmosphere. This also is true to an extent with dryers. If a fan is at work drawing cold air through a stream of descending wheat, and that air is charged with dampness, then harm may be done instead of good. On the other hand, it does not necessarily follow that the air is in the same condition when it is simply raining, and there is no doubt that of the two the latter is the more favourable time so far as this operation is concerned. The wheat also will show this to be true, for it will be found to have acquired a cold, clammy feel if the barometer shows a low temperature, such as indicates this state of things generally, and it would be always wise to suspend as much of the operation as possible pending a better state of atmospheric surroundings.

In closing this chapter on wheat washing and conditioning, it is calculated to be of future benefit to milling
students that a synopsis of the whole be included in the
form of a set of questions bearing on the chief points
contained therein, and which all interested millers will do
well to bear in mind and work out from their experience,
aided, perhaps, by the efforts put forth to guide them in
a right direction towards the object in view.

1. What is the desired end to be reached in the cleaning or preparing wheat for the making of flour ?

2. Give a list of machines which operate upon the
   wheat previous to being stored in silos, and say, if
   a travelling band runs 500 ft. per minute, what
   law prohibits its running at even a greater speed ?

3. Describe the various milling and warehouse separators, and state wherein they differ in structure
   and in the work performed as usually employed
   in both large and small mills.

4. What are the objects sought to be attained by
   grading Wheat for barley, cockle and rye cylinders ?
   Describe the action of these machines and the
   difference in their internal structural arrange
   ments.

5. Scourers and brushes.—What is the difference in
   the two types as regards power absorbed, efficiency
   in working, wear and tear comparatively, and the
   reason for any preference ?

6. Dust collectors for screen-room purposes are mostly
   of one type in principle—that is non-textile.
   Describe the circulation and deposition of the
   different components forming the substance
   exhausted from the different machines on a dry
   system of cleaning.

7. What are the main objects sought to be attained
   in the washing of wheat, and how is the wheat

prepared for the process—that is, what machines does it pass through previous to entering the washing tank?

8. What process immediately follows the immersion of wheat in water, and for what purpose?

9. Steam or hot air is used as a means of correcting or conditioning wheat which otherwise is not suitable. What is chiefly to be guarded against during the operation, first as regards the wheat itself, and secondly as regards its after condition and subsequent treatment?

10. After the hot air process wheat is scoured and brushed. Give a detailed answer on the effect sought to be produced at this stage, and the reason why it is considered of importance.

11. Following upon this, coolers or dryers are used in immediate succession to the last operation. Why should it be thought necessary to lose no time between the operation of heating and cooling?

12. Automatic weighers are in use during the cleaning process of wheat. State where they are placed, and for what object. Why is one not considered sufficient?

13. How many degrees of heat (Fahr.) is it considered safe to use without risking injury to the vital principle of wheat in a strong foreign variety and on a weaker brittle variety respectively, as, for instance, strong Russian and Indian.

14. Storing for longer or shorter periods is recommended after the whole treatment as given in this chapter. What end is sought in thus mixing the various sorts together before being reduced by chilled iron rolls?

15. Name where air currents can be beneficially employed otherwise than upon the machines mentioned, and for what purpose.

16. What is separation by magnetism, and where does the separation take place?

17. Should impurities be present in the wheat after a dry process of cleaning, what are they most likely to be, and what will be their effect on the flour?

18. Is it customary to wash English or other so-called soft variety, or if not in a condition requiring it, what part of the wheat cleaning system is best adapted to correct excessive natural moisture?

19. Has the number of revolutions of any machine or other appliance connected therewith sufficient influence over any operation to alter the result which should accrue from the passage of wheat through the machine? If so, in what way?

CHAPTER V.

The Break System.

In the early days of roller milling, and even to within a comparatively recent period, great diversity of opinion was manifest among milling engineers as to the precise mode of operation necessary to give the best results on each successive roll. The question of crease dirt and the careful splitting of the berry longitudinally across this crease, and the good results obtained thereby, invested the first break roll with an importance all its own. Even now there are, perhaps, some mills still intent on the practice, although, no doubt, with a different purpose than was originally intended. Very likely in the majority of cases a solution will be found in the fact that notwithstanding all that has been written against it in the light of later and better information, it is still utilized as a sort of finishing roller to the wheat cleaners, and as such is a valuable auxiliary.

To begin this series, I propose to submit a plant working on four breaks. As a rule five breaks are as short a system as is general with moderate and large mills. Some, indeed, having six breaks are not altogether unknown ; but, as in all innovations, extremes are tried until a medium course shows itself as best adapted to the highest results. In explanation of the technical terms applied at this stage, it must first of all be stated that the number and size of both break and reduction rolls is calculated at so many inches per sack of flour made in an hour. Rolls up to 36 inches in length are usually 9 inches in diameter —except for very small rolls—and above that are rolls of 40, 50, and 60 inches, whose diameters are 10 inches.

The basis of calculation is usually on the length, and a fair average length is now reckoned at 35 inches per sack on the break rolls. Some engineers give rather more and others rather less. Thus if there are 14 pairs in a run of four breaks and they are all 30 inches by 9 inches, it will be worked out like this :—

```
            30 inches (length of rolls)
            14 (number of pairs)
           120
            30
Sacks per hour, 12 ) 420 ( 35 inches per sack.
            36
           ----
            60
            60
```

The corrugations call for a word of explanation. It is usual to have the wheat divided just before it gets to the first break roll, that is to say, the large grains are sent to one pair and the small to another. This ensures a greater uniformity of size for the next or after treatment.

Diversity of opinion raged for a season as to the best form of corrugation suitable to English wheat mixtures, but the consensus now points almost exclusively to the so-called sharp saw-tooth fluting as furnishing the best all-round results. Of these there will be eight to an inch for the large wheat and 10 for the small wheat. The corrugations are not cut parallel, but at a slight angle. This helps the shearing action when at work, and also prevents the rolls from catching each other or interlocking, if by any mischance they are set too near. The rolls run at different speeds to accelerate the separation of the branny coating and to prevent the kernel from being flattened.

The following is a table of speeds and corrugations on a four-break system :—

		Fast.	Slow.
		—————SPEEDS—————	
1st break 8 and 10 corrugations per inch ...		400 revs.	160 revs.
2nd ,, 14 ,, ,,	...	400 ,,	160 ,,
3rd ,, 18 ,, ..	...	400 ,,	160 ,,
4th ,, 26 ,, ,,	...	480 ,,	160 ,,

Robinson Diagonal Roller Mill.

An innovation in the relative position of rolls has been introduced by Messrs. T. Robinson and Son, Ld., Rochdale. The rolls are fixed diagonally.

The most striking features in this machine are its capacity and its cool and even grinding.

It differs in construction from the ordinary type of roller mill in the following points :—

    1. The relative positions of the fast and slow rolls are such as to give better access to the feed above the rolls, as well as to the scrapers and stock after grinding.

    2. The difference in the diameters of the fast and slow rolls permits of their running at the same speed, and yet obtain the correct differential at the point of contact.

Other noticeable points in the machine are its compactness, solidity and general handiness in working.

It will be noted that the differential is obtained by the difference in the roll diameter, one roll being 13 inches and the other 10 inches.

Other makers of diagonal machines include Messrs. Briddon and Fowler, Manchester, whose " Britannia " rolls are highly appreciated by those who have them in daily use.

It is of prime importance that rolls should be set in perfect parallelism, and the first thing to do is to see that the fixed roll is correct in the framework, i.e., both level and square. Once this is done, it is then easy to adjust the movable roll with the aid of the face plate. It is sometimes the case that rolls are grinding unequally at opposite ends—one end is too close and the other not close enough. As a rule, the correct way of rectifying them is to open the end which is too close, and so relieve the tension, and it will usually be found in practice that the opposite end is ready to respond by springing a bit closer to proper working order—in fact, the roll will get parallel. If a smooth roll is much out in this way it begins to give evidence by a sound more or less audible. The end of a roll out of line will also get hot, and this is a very sure guide to their correct alignment. To obtain a proper finish on break

rolls care should be exercised to make the first and second do as much as possible, without, of course, overdoing it. If this is not seen to the latter breaks are overcrowded, and a good finish of the bran becomes an impossibility.

The best semolina also is only possible on the coarser corrugations, and if the opportunity be missed here the result is a large quantity of break flour and dirty middlings, with only a small quantity of the larger particles, a condition of things to be avoided where a good standard product is expected.

If the initial rolls are properly set the cleaning of the bran is not a difficult matter, but care will be needed to see that they (the bran rolls) are doing their proper duty. Many cases are to be seen where the bran, having been cut up on the second or third break rolls, comes in a very unsatisfactory state to the cleaning or bran rolls, and if a man opens these latter a little so as to avoid further cutting this also is a mistake. Really the making of the bran depends mostly upon the first three rolls more than upon what are called bran rolls. If the bran comes to the last break thick and hard it is a useless task trying to make a satisfactory job of it. The best thing is to go over the whole lot step by step, and arrange the adjustments, and if this be done with a knowledge of what is required there will be no difficulty in the right apportioning of the work, and the finish will be what is wanted, so far as regards the bran. Bran rolls, of course, are expected to have something to do, and should be set up to grind. They should always be set close enough to catch any unfinished material which comes on. If they are grinding, the product will be more or less dusty; but if not, it will be exactly like the feed, and in passing this is an easy rule to work upon, as too little is often done here, and the bran feels hard, although it may look clean, but the percentage of flour will be proportionately less. This, again, does not mean overgrinding, and a great deal of mischief may be

caused in this way, as bran powder is by far the most
pernicious agent there is in spoiling the colour and flavour
of flour.   It will be mostly found that either over or under

Dell's Roller Mill.

grinding is caused by unequal adjustment or distribution
of the work required to be done.   It is not often that bran
rolls, or, in fact, any pair of the series are set up bodily
too close, or, if so, not for long ; but greater attention is

more necessary to carefully examine if both ends are set in exactly the same position. Especially in the case of bran rolls one end is liable to get up closer than the other under the hands of any but a skilled rollerman, and consequently while one-half, or not quite one-half, of the roll is making bran powder, the other end is quietly slipping it through practically untouched. The proper setting of bran rolls requires practice in seemingly greater proportion than at first sight appears necessary ; but it must be remembered that in finishing up this product less and less duty is required in so far as concerns quantity, and more and more skill as regards quality.

If the bran be very small there is direct evidence of severe treatment on one or more of the earlier breaks ; if it be curled, too much work is being thrown on the bran-cleaning rolls. In both cases flour is made, and that of inferior quality, the stock for purification is degraded, and the patent flour is lowered in quality through being made from inferior middlings. The straight run is discoloured, and the net result is to place the finished product at a lower intrinsic value than would be the case if all the break rolls were given their proper amount of work to do.

Sometimes bran shows the impression of the corrugations ; but this also should be avoided. It is bad enough when it arises from the last break ; but if it emanates from any previous one it means that it will be more or less cut into strips, and to some extent spoil its marketable value, besides consuming more power than is necessary, and increasing the quantity of bran middlings.

Rolls scrape or tear both sides of the bran, and when it is considered that the skin of wheat is about 1-400th part of an inch in thickness, we ought to form an opinion as to the skill required to successfully pilot this product through the necessary operations.

It is an easy matter to locate the mischief when the bran is being snipped, because the marks will correspond to the number of corrugations on the guilty roll. Science also tells us that bran, if too heavily pressed, yields a moisture which is really a dye, and by so much will spoil the later

Armfield's Roller Mill.—J. J. Armfield and Co., Ringwood, Hants.

runs of flour. The skin of the wheat is, as we have already learnt, very susceptible to heat and moisture, and both are caused by overcrowding the feed on the latter break rolls, so that as much exhaust as is possible ought to be laid on to keep the stock cool, and thus avoid condensation, chokes, and consequent irregular flow of material.

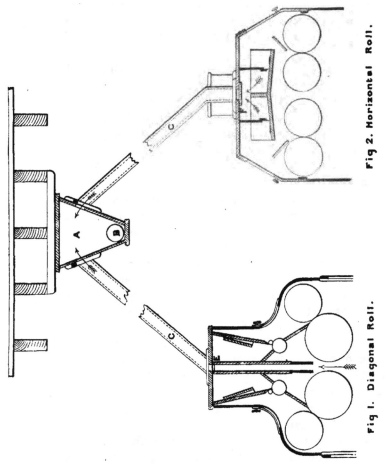

Fig 2. Horizontal Roll.

Fig 1. Diagonal Roll.

Robinson's Exhaust System.

Broad, clean bran is generally an indication of intelligence and skill in the handling of the break rolls, and also an indication of clear flour and clean feed.   It is evidence that pulverised particles have not been distributed among the various grades of middlings.   It may be sometimes expedient to give the bran this extra squeezing for financial reasons ; but whenever and wherever it is practised the after product ought never to join a respectable sample of flour.

It may be stated here that there are three well-defined rules which explain the object of breaking down wheat, and they are : 1st, To get the bran away from the inner or floury portion of the wheat in as near a perfect state as possible.   2nd, To make as large a percentage of purifier stock as possible on every break roll ; and, 3rd, To make as small a quantity of break flour as is consistent with the fulfilment of the first two rules.

Some mills have as much as 40 ins. break roll surface per sack per hour, and where the highest results are reached, and all other things a secondary consideration, there is no doubt it is by far the best plan to work upon, and in the long run will be found to pay best.   One simple reason for this is that while more power is absorbed in driving an extra pair of rolls, it is almost, or quite, atoned for by the fact that all the breaks are working at less pressure with thinner feeds than would otherwise be the case.   Rolls will certainly last longer if not overworked in this manner.   The feed ought to be regulated so that the particles under operation are touched by both rolls at the point of contact.

If, on the other hand, the feed is so thick that a grain or part of a grain is crushed between one roll and a cushion of other grain, then there will be multiplication of break flour and also of misshaped middlings.   This fact is not enough reckoned upon when short systems are employed, nor does it form a point when calculating speeds, peripheral or differential, spiral twist and number of corrugations, and, therefore, to force any roll beyond the single layer or thin stream is against all the primary laws of gradual reduction.

The spiral twist given to the track of the grooves will average about $\frac{1}{4}$ in. to the foot length, except for bran cleaning rolls, when it is usual, or becoming more so, to increase it to as much as $\frac{1}{2}$ in. The generality of roller mills have a differential speed of $2\frac{1}{2}$ to 1 on the first three breaks, and 3 to 1 on the last. This greater difference in the speed is to ensure a slightly keener shearing action on the bran flakes, and so scrape, as it were, the surfaces without so much pressure being employed. The grooves on the earlier break rolls should be kept sharp and fairly deep, so as to be a sort of momentary refuge for the semolina just dislodged by the process. If too shallow broken pieces are inclined to get crushed, and by this means interfere with the good results otherwise to be obtained.

The break rolls must be looked at as a whole. Whether three, four, or five operations are employed the work must be distributed discriminately. Too much must not be left to the last machines, for if so there will be a mass of fine stuff to deal with which will almost defy separation in a satisfactory manner, the bran is hashed up into a condition almost unsaleable, the scrapings are spouted to some of the latter reductions in quantity too large to be properly dealt with, and a bad finish and poor percentage results. It would astonish one to figure out what 1 per cent. difference let slip in this way means in the course of a year. What is looked for in a man who has charge of this department is not merely one who can get all the flour off the bran during the break process, but one who has a technical knowledge of the whole process. It really is not a difficult matter to get a good percentage; but the flour does not give that satisfaction to customers it otherwise would if greater knowledge or skill was bestowed on its manufacture. A lack of this knowledge will show itself also in the care of

D

Four Roller Mill.—Henry Simon, Ld., Manchester.

the machines themselves. Non-attention to the regular adjustment of rolls which are slightly out of parallel, uneven wear of journals and pins, unpacked brasses, and almost a hundred other little items which are apt to slip the notice of those not up to date in technical knowledge, or careless in the exercise of it, all tend to lesson the standard of the finished flour, and when a stoppage must be made the overhaul is more expensive than would have been the case had the numberless little things been attended to when as yet they were but little things.

A roll that has stood the test of over 20 years' work is that of Messrs. H. Simon, Ld., of Manchester.

The design is compact and neat, the construction solid and durable, and the fittings and adjustments simple and accurate.

The frame of the machine is composed entirely of iron and is absolutely rigid.

The rollers are made of specially hard and tough chilled iron with strong steel axles.

The journals work in heavy phosphor-bronze bearings so arranged that all wear is taken up automatically. The lubrication is also automatic and continuous.

By means of hand levers the rolls can be instantaneously set to work or thrown open, the same movement automatically starting or stopping the feed.

Both ends of the lower rolls can be adjusted accurately and independently of each other.

A positive differential motion is secured by means of double helical teeth wheels running in oil.

Each pair of rolls is provided with an automatic and adjustable feed gate while by special devices provision is made against any condensation on the feed plates or within the roller frame.

All parts of the roller mill are easily accessible.

D 2

Milling machinery in the roller stage is calculated upon a certain duty basis, pretty accurate, but not much allowed in the way of margin, and no opportunity must be lost to maintain each section at its highest efficiency. The middlings from a well-regulated line of break rolls should be of the size and shape best calculated to please the intermediate run of sieves, graders, and purifiers, that these machines in their turn may deliver them to the smooth rolls in a fit state for immediate conversion into flour granules. Any remissness will result in augmenting the quantity of break flour, and that means more or less dirt and pulverised branny fibre in the flour sacks and less amount of stock for purifiers; and they in their turn will sift out "bare" and deliver their product in less quantity and also in secondary condition to the reducing rolls. If the fault be aggravated the smooth rolls get short of feed and take it out of what they do get by extra pressure, the ultimate consequence being flour more or less slippery to handle, because in reality the life has been crushed out of it. This is, of course, taking an extreme case; but with ordinary care and attention little lapses from time to time will be noticed all over the mill, which can usually be corrected before any mischief is done if given right treatment at once.

Another machine of note is the roller mill of Messrs. Amme, Giesecke and Konegen.

The frame of this roll is built on a solid casting. The four sides of the upper structure are bolted and rivetted on the foundation, making a frame of exceptional strength and rigidity. The facings and borings on the frame are all done in such a way that they stand parallel and perpendicular to each other. The roller bearings are fitted with loose ring oilers running over the middle, and a feature distinguishing this mill is that the bearings are adjustable sideways for taking up the slack caused by the wear and tear on the shoulders of the brasses. This prevents the rollers working sideways.

The mill is fitted with double roller feed and automatic feed valves. The adjustment of the movable rolls is so arranged that there is one adjustment for each side, a second adjustment to move them simultaneously, and a third is provided for the stopping and re-starting of the roll. By moving a lever the grinding and feed rolls are thrown out together, and, by the same process, re-started for work. Each side of the movable rolls is separately adjustable, thereby keeping both working rollers constantly parallel.

The handling of a line of break rolls is very different in different mills. For instance, there are many mills working on foreign wheats entirely; others are using only a proportion—great or small—while in an extreme case it is possible to come across those who are milling entire samples of English. As is well known among all practical men, the working of these different mixtures requires skill. What would be right for any one would be unsuitable for the remainder, and it is even possible that two samples of the so-called same wheat will show considerable variation of behaviour when passing through the rollerman's hands. Take, as an example, a small mill grinding a parcel of home-grown wheats. We will not here enter into an analysis further than to say that all our English wheats come under the denomination of being soft. Times may happen in very favourable seasons when it will approach some of the milder foreign sorts; but generally it is designated a soft wheat. Now, a full feed of this passing through the rolls requires very skilful handling if anything like a good result is to accrue from the process. It would require very gentle treatment if the aim were to get as many middlings as possible for purification, and a greater differential would be of much benefit. The reason for this seems to be that the cells composing the endosperm, flour, or kernel, as it is variously called, are not built up so compact as others to be named, consequently great pressure is not required in breaking them apart to extract middlings of suitable size and to shear the bran away in

Four-Roller Mill.—Amme, Giesecke and Konegen, 59, Mark Lane, London, E.C

large flakes. Granted that a sufficient surface area was available, the task of breaking down English wheat would not be a very difficult one in an ordinary season if the making of middlings was not the principal object aimed at

Taking the other extreme, that is a hard foreign wheat in the majority of instances, were it not conditioned in some way, every break roll would be answerable for making bran powder, the wheat in its natural state being so dry and short that decent bran would be almost impossible. However, we will suppose it has passed through the machinery named and described at an earlier period, and that for that purpose it is as perfect as milling science can make it. It is perfectly clean, the outer integument has been softened, and thus made more tough to facilitate its removal by the action of break rolls ; but remember this conditioning process only applies to the bran really ; the inner portion is practically unchanged in its nature, and remains a hard, flinty mass, which a good nip by a break roll disintegrates or splits up into what we call middlings. Some of our milling engineers tell us that the whole berry is affected and changed in texture, but we will not stop to argue here, as the primary object of all conditioning was, and is, to render more certain of accomplishment the removal of the bran while making the largest amount of semolina or middlings possible, leaving everything else of importance to a later stage. The thinner skin of the variety under treatment renders it a matter of delicacy in adjusting rolls so that the shearing action shall be minimised, and that examination will show a large quantity of broken particles of nearly the same size, clean and bright, as little adhering flour as can be managed, and bran as large as is compatible with the just mentioned process. If the feed and condition of the wheat be regular, the desired end to be reached is by no means impossible of attainment.

Far different is it when a mixed wheat has to be operated upon (that is a proportion of hard and soft sorts combined in a mixture), for then everything depends upon its condition or the treatment it has previously undergone. It is the custom in mills which do not wash wheats to make a blend a few days in advance of requirements and let the mixed quantities lie together to assimilate, which means that those kinds containing moisture to excess will when in contact with hard dry varieties, part with some of it, while the hard wheats will pick it up, and each bring themselves more into line by these opposite tactics. If time can be spared, this old established custom works to perfection, but should it be desired to grind the mixture at once, then the rollerman has very difficult work in handling wheats so varied in texture and build on one and the same roll at the same time. As has been stated just now, the pressure and differential are not alike for both wheats treated by themselves, so that it will be clearly seen that to break up hard and soft wheat mixed together is an operation that cannot be performed with satisfaction. What is sufficient for a hard, short-grained wheat is not so for a soft, tough variety needing a more extended shearing or scraping action in order to free the grain particles without making too much flour ; as a matter of fact the feat is never performed with the good results which would naturally follow were a better distinction drawn between the various components of the wheat mixture, by making sure that before they are operated upon for flour-making purposes they shall, as far as possible, be in the most suitable condition as regards themselves for supplying the maximum amount of middlings at the expense of a minimum amount of break roll flour.

The moving parts of the break rolls must run freely, quietly, regularly, and without heat and undue wear and tear. Lubrication should be frequently attended to, and spent oil, grease and dirt removed before there be any accumulation. Attention must be given to belts, and as occasion requires laces and joints inspected.

Regarding roll feeds the one illustrated is a good example of what is provided.

It consists of two feed rollers, arranged one above the

other, the bottom one—which is larger in diameter—runs at the same speed as an ordinary feed roller; whilst the upper one—which is much smaller—runs slower. The small roller acts as an ordinary feed gate, and is supported in its bearings by levers connected to a moveable feed board in the hopper, thus the distance between the two feed rolls varies in proportion to the quantity of feed coming on to the machine.

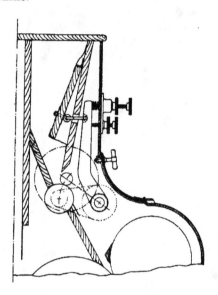

Robinson's Twin Roller Feed.

Examine the corrugations regularly and often, work adjustable parts to keep up the efficiency of the rolls; keep a good look out to the feed of each roll, that it is free from dust, regular in quantity, and otherwise up to the required standard in quality; note whether enough or too much exhaust is being applied; work each individual pair so that they perform their due share as far as possible, not getting in front or behind as to the duty expected of them. For a

Briddon and Fowler's Roller Mill.—Briddon and Fowler, Manchester.

description of the various parts forming a vertical roller
mill a typical pattern is furnished in almost every town in
England.

Horizontal rolls are the other type it may not be amiss
to say a word about in regard to construction. While
possessing all the qualifications of the vertical kind in
regard to the engineering, rigidity, soundness of the material
employed in construction, they differ in that the rolls are
placed in an horizontal position regarding each other, and
the wheat is fed between them, the adjustable arrangements thus working in a sort of slide which on the face of it
looks as if a more delicate handling was possible than when
raising or lowering, as is the case with vertical rolls. However that may be must be left to individual opinion, but
private experience suggests that the horizontal type
more than hold their own in the department connected
with the breaking down of the wheat, that is, the break
system. There is not a great amount of pressure required,
they are not liable to get out of order, and as a rule stay
where they are set, and are to be relied upon in every way.
A great deal has been written about the absence of friction
in vertical rolls, when the top roll is made the adjustable
one. That, of course, is granted so far as that roll is concerned; but then the lower one has to withstand a strong
downward thrust in addition to the friction of its own
weight, whereas in the horizontal type a side pressure only
is encountered, so that whatever other advantages may be
claimed, it is evident that some drawbacks are present in
both, and the matter is better left alone for individual
inspection when all other things are taken into account by
the prospective user. In break rolls the concussion caused
by the disintegration of the wheat berry is more or less
violent, and lends itself by vibration to the loosening of
screws connected with the means of adjustment, and with
vertical rolls this allows for gradual lowering until the
operation becomes a grind. Once let these adjustment

screws get loose and gravitation does the rest. The springs with which both types are fitted act, and are intended to act, as a cushion in case any hard foreign substance, such as a nail, screw, elevator nut or bolt finds its way past the roll hopper. Were it not so, these things would almost certainly cause the breaking of a spindle—perhaps two. They are made stout enough to resist the pressure ordinarily employed and to act only in emergency. Vertical rolls certainly have an advantage in not taking up so much room as their rivals, and the operation is more open to examination by the workman.

Summarising what has already been written in connection with the breaking down of the wheat, the points primarily important for constant attention and supervision may be condensed into the following remarks :—

1. To manipulate each pair of rolls so that they are adjusted parallel to each other, and in the same plane.

2. To feed them evenly the full width.

3. To keep the pressure even and sufficient at each end to do the work required of them.

4. To be sure that the proper differential is maintained on belt driven rolls, according to requirements.

5. To attend to lubrication, and keep all moving parts clean and free from accumulation of every kind.

6. To have enough exhaust applied to draw away all hot air and light tissue, and so prevent condensation or sweating, which, if not done, interferes with the machines which immediately follow, and in so doing reacts on the material under treatment, and that to its detriment.

7. To immediately report when rolls require refluting which should be at least once a year, and, perhaps

earlier, if they are constantly at work on wheats which are tough and depend for disintegration mainly upon a quick shearing action.

8. To keep belts in thorough repair.

9. To frequently test the product from each break roll.

10. To see that the bran is well cleaned, but not cut up or abraded.

11. To notice the feed coming to each roll, whether dusty, irregular, or of the right quality.

12. To keep the whole of the breaks uniform as to quantity and quality of work done upon each pair.

13. To pay attention to the finished products.

14. To make the highest percentage of purifier stock.

15. To make least percentage of break flour consistent with the best total results.

16. To run all machines evenly so as to avoid undue wear and tear.

17. To notice any and all variation in the wheat coming to the rolls, whether relating to the raw material, its condition, atmospheric changes, or arising from any other source.

18. To occasionally test main shaft speeds and all products to enable one to keep the mill going regular, and so avoid loss of time and capacity.

19. To frequently run off the residue from the spent oil chambers, and so prevent undue friction from the clogging action which is caused by returning the same oil along the spindles beyond the limits of its lubricating power.

20. To inspect toothed gearing occasionally, so that vibration is minimised by having them trimmed as soon as signs of wear begin to show.

21. In stopping and restarting break rolls see that there is sufficient feed in the stock hopper to prevent them running empty.

The cutting edge of the fast-running roll is set forward and the slow runner backward.   There has been much controversy and experimenting with regard to the proper setting and cutting action, and some experts suggest even now that the slow roll ought to be turned round, but to the ordinary mind it is a very simple matter.   We either want the cutting action or we do not, and it has been abundantly proved it is the best for English wheat mixtures, so that further discussion upon the supposed merits of dull rolls would be entirely useless.

It has often been objected that feed rolls travel too slow, and that they do not deliver the feed to the roll in as thin a stream as possible.   While not going so far as to say this is wrong, I would point out that there is a limit to this in the amount of feed coming to the hopper, and also that the feed dropping upon a larger surface area gets spread out considerably before entering the grinding point of contact, and also that too high a speed of feed rolls would also result in disintegration at the feed gate owing to increased friction there, so that, on the whole, it is best not to exceed in this, at the head of the mill especially. But this mostly applies to smooth rolls.

I cannot close the chapter without calling attention to several articles which have appeared in the milling press, advocating a return to a shorter method of producing flour by a different adjustment of the break rolls.   It is argued that wheat is not the same since washing and conditioning have become general, and that the gentle treatment we are now supposed to employ in getting away the bran before making flour, and the production of a large amount of middlings for purification, is not of the importance it was under different conditions, and that, in fact, a shorter cut should be made by having rolls of finer corrugations and

higher speeds—differential and periphery—to shear the
bran away in large flakes, and at the same time make 40
per cent. or 50 per cent. of flour.   The argument is based
mainly upon the fact that wheat is now absolutely clean
and in condition, which greatly favours a more direct action.
What has to be guarded against is the production of bran
powder, and if wheat is in that perfect condition specified
bran powder is not at all likely to be made.   Without
taking up a controversial attitude, I should like in a few
words to put the case as it now stands.   It has been shown
in these pages on breaks how step by step we move carefully in order to get the largest amount of middlings, to
avoid cutting up the bran, the making of flour, and the
production of very small powdery particles believed to be
the fibrous material composing the outer skin of the wheat.
This principle of gradual reduction was first enjoined mainly
because crease dirt was supposed to be present—that is,
that small particles of earthy matter were lodged in the
longitudinal crease of the berry.   In the light of further
experience this was proved to be a fallacy, as nature is too
complete in her handiwork to allow anything of that sort.
However, it was found that the dried up, shrivelled life
cord or placenta was the substitute for the discarded crease
dirt theory, and therefore the gentleness on the breaks was
still continued.   By-and-bye washing became more general,
and then conditioners made their appearance, so that the
continual adding to already existing machinery made
millers cast about for some compensation in the way of
treatment, and the idea was launched that since the
original notions of gradual reduction by break rolls had
undergone so many modifications the reason for its con-
tinuance did not exist.   It should have been stated that
after the placenta theory wore itself out bran powder was
mooted, and it was found that wheat in a dry state is ever
parting with its outer shell or coat in the form of dusty
fibre ; but even that is obviated by conditioning, hence the

comparative prominence which the latest innovation has been raised to. It may be stated that in America there are two distinct milling systems—viz., flour millers and middlings millers, and needless to say the latter are those using the hard dry wheat of the great North-western Provinces, while the flour millers are chiefly occupied in converting the mild winter varieties by means of such a short system as is here sketched out, and it is this fact which English millers have apparently laid hold of, seeing that well conditioned wheat to be perfect in every way will exactly resemble winter American in so far as the latter's characteristics are concerned.

In the short or direct system the bulk of the flour is made on the first two breaks, and in such a way that the impurities let loose are not so finely reduced as when exposed to repeated operations, so that as a consequence they are the more readily separated from the flour. It would appear at first sight that the innovation was a great advantage, and this may be so where only a moderately high-class production was wanted; but in the face of the present home and foreign competition it is more than likely that the great majority of millers will stand by the middlings-making system until some more tangible proposition is put forth. It may be that a direct run like the one mentioned would make a good straight grade flour, but the value of the residue would prevent any high-class patent being manufactured, as the middlings sent for purification would undoubtedly be much inferior at all points to what is the case in the daily run of the prevailing system. That is to say, that the highest results can now be obtained, but under other conditions, such as, for instance, the sudden "death" system, only a moderate degree of perfection is possible.

A list of questions bearing on the subject matter connected with the breaking down of the wheat berry is here inserted, as calculated to assist millers in grasping the main points aimed at during the operation :—

1. What is meant by the "condition" of wheat, and how does variation affect the breaks and also the products?

2. What is the usual differential speed as between the fast and slow rolls, and how does more or less differential act on the broken wheat?

3. What are the main objects sought to be attained on the break system, and how is it best to secure the desired ends?

4. What is the cause of bad colour in flour made by the break rolls, and having found it, what is the best remedy, apart from a short system?

5. What is about the angle at which corrugations are cut, or, what is the usual distance from the horizontal in rolls 18 in., 24 in., and 30-in. in length?

6. Describe the movable and fixed parts of a pair of rolls?

7. What is the benefit derived from using exhaust on rolls, and how is it applied?

8. Would the exhaust be better applied to the material before or after passing through the rolls, give the reasons for same?

9. What is it requisite to do regularly and often to keep rolls up to the correct distances?

10. Give one or two ways of knowing when break rolls are getting dull, and, when examined, what particular point is prominent for inspection?

11. What is the result of under and over feeding break rolls?

12. What is a feed roll, how does it act, and what is its chief duty in delivering stock to the rolls?

13. What prevents rolls from going together, and what provision is made for foreign substances which by accident get into the stock hopper ?

14. What is the effect of dull break rolls upon the middlings, break flour, and bran ?

15. What is about the usual surface area allowed on break rolls per sack of flour manufactured ?

16. If the first, second, or third break was being overworked, what would be the result ? Is it possible by examining the finished bran to locate anything being wrong ?

17. Why is it necessary to give larger surface area per sack per hour on certain breaks more than on others, and what is the condition of the stock that makes it incumbent to do so ?

18. Give reasons for or against the different designs of rolls used for the break process, *i.e.*, vertical, horizontal and diagonal ?

CHAPTER VI.

SCALPING AND GRADING.

The principle of the gentle treatment of all break stock is now fully established in all parts of the country. Following upon the centrifugal came the round and hexagon reels, but they have now been largely superseded, and the most prevalent way of treating the middlings in their raw state is relegated to sieves. The principle of action is as follows :—From each break roll the product is conveyed to sieves, which are usually double, that is to say, one sieve has two surfaces, a primary one and a secondary one, the latter being underneath the former. The top sieve receives the break product in bulk, and the mesh being somewhat coarse—18 or 20 wire, or zinc equal to that—allows all material less in size than the perforation to drop on to the under sieve, and what tails over proceeds to the next break roll. Of the sifted product, the larger sized material is retained for immediate entry to the coarse purifier, and the throughs are dusted free from flour on a centrifugal usually, and purified as middlings. The operation of scalping is simplicity itself, in regard to the mechanical part of it, and if a mill keeps pretty generally on the same class of wheat, and enough surface per sack is allowed 'on the sieves themselves, a very little attention to their moving parts suffices to keep them in good working order. The danger is not that of themselves, but rather in what quantity and condition stock is spouted to them from pre-

vious machines. As a rule they are erected with an
incline of about one inch to the foot, and are fitted into
an adjustable eccentric having a variation of a little more
than an inch. They are hung on rods or leather straps,
which can be regulated in length by buckles or other
device, and are run at a speed of from 180 to 200 revolutions per minute. Ordinarily they have a balance pulley
to subdue or prevent vibration, and the inlet and outlet
spouts are connected to the machine framework by some
kind of flexible material to prevent unnecessary dust and
waste. They are best with independent drives from a line
shaft, so that if anything goes wrong it is confined to that
particular spot, and does not upset or interfere with the
temporary efficiency of other accompanying machines.

Scalping is simply a separation of coarse material from
fine, and must always be considered in relation to the
break system of a mill. In this sequence it is used to
separate the branny part of the material, which needs further treatment on the break rolls, from the semolina middlings, dunst, and flour, which get their further treatment on
purifiers and centrifugals further down the system. At
the time when roller milling may be said to have commenced in this country, some 22 years ago, there were two
systems of scalping in operation, namely, by reels and
centrifugals. Each had its patrons, and amongst them
were numbered the ablest millers in England. The
advocates of each have now changed their opinions for
the most part, yet when we consider the matter there
seems much to be said on both sides. The opponents
of centrifugal scalping argued that the action was too
severe, and that it had a detrimental effect on the semolina
and middlings of the break chop, and that a reel, by its
more gentle motion, did better work. Theoretically, no
doubt, this was so, but the centrifugal was made very
short and got rid of the stock quickly, while on the other
hand the reel was in comparison very long, and retained the

feed much longer, so that in reality there was little or no practical difference in the quality of the work performed. Then the " rotary " appeared, and immediately caught on. On its introduction it was rough, but, as its friends said, it was also ready, and in its improved form it gives satisfaction, though whether it will continue to do so in face of the re-introduction of the plansifter remains to be seen.

As a successor to hexagon and octagon reels, and as a competitor with rotating sieves, there has been introduced a sort of hybrid machine called an inter-elevator reel. This machine is round, and rotates rather faster than was before usual, and it is fitted inside with lifters, by which means the material fed into it is distributed over a much larger area of surface than was previously found practicable. The capacity also is greater—nearly double—and its action is relatively more gentle than its former prototype. We are told by those who have had considerable experience with them that they make a first-class scalper. It is probable that along with sieves they divide the honour. There are many good points in both. Both are simple, and not liable to get out of order, and will last a lifetime ; the parts of each are so easily removable that in practice they are not worth making note of. To my mind, however, inter-elevator reels are at their best work in the next process of grading the separated chop. They are an admirable improvement on the old style of reel, and have a greater capacity, and there is not the liability to disintegrate the unpurified particles to the extent which now obtains in many mills. There would seem to be just enough of the friction necessary to separate partly-loosened flour particles without scouring the rest to any objectionable extent. Some force is imperative to accomplish what the sieves have left in a sort of half-and-half condition, and this is just what is aimed at. The original slow-running chop reel had not got enough about it to finish its work, and must needs send the most difficult portion to a

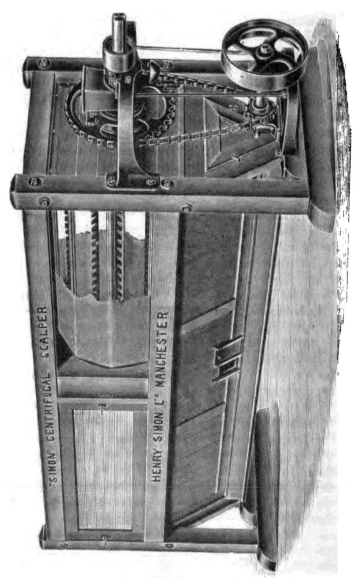

Simon's Centrifugal Scalper.

dusting centrifugal, and in some instances this machine overdoes itself in attempting to whip loose partly-adhering substances so as to fit the residue for the purifiers. The inter-elevator reel steps in here, and combines the merits of both without partaking of their faults, and would seem to give satisfaction on almost any kind of stock. Its travel is regulated by the machine's momentum without the same regard to its product's specific gravity. It is a continual slide through in place of the drop and tumble of its predecessors, and it must at almost first sight catch the eye of millers for its surface qualifications.

Going back to the beginning, the first desideratum of roller milling at this stage is middlings—large, medium, and small—as many of the first as the rollerman's knowledge can produce in handling the break rolls, and as few of the finer particles as is consistent with the best results. Too many fine middlings at first hand, as it were, means a larger dusting surface of silk, and that, with more or less harsher treatment, also means break flour in the proportion that smaller middlings are made. Middlings are what the break rolls live *for* and what the purifiers live *on*, and the highest duty of all dressing machines is, after getting them perfectly pure, to hand them over in the shape of flour, and in the same condition of purity, to the flour packer. It is usual to separate the product of scalpers and graders into three or four divisions. The introduction of the plansifter opens up a much larger sphere of possibilities in this department of grading, and reminds us how careful Continental millers are in this respect in classifying their middlings. What labour, divisions and sub-divisions they subject the stock to, and we must confess that the result is highly satisfactory. It is not intended for anyone to infer that the same tactics should be resorted to in England, because many things make it impossible, but the favour this latest machine is enjoying opens up greater possibilities in grading than

what are known or practised in the generality of roller mills
at work, and opinions shadowed forth by thinking men
point in the direction of getting as many sizes of
middlings as are present, getting them separately, and
treating them so on the purifiers. These latter machines will
also lend themselves to this altered condition of things
readily, and the best re-adjustment which suggests itself is
to arrange the sieve surfaces in narrow strips longitudinally
instead of laterally as now. To illustrate this meaning we
will make a case. A purifier, 60 in. by 30 in., is (say)
treating stock from a so-called grader ranging from flour
down to No. 8 silk, that is after being dusted. If the
purifier was partitioned the full length into 6-in. spaces it
would take this stock divided up into five sizes, and for
after purposes this is surely a better way of grading than
the other. As these machines will perform this they are
well adapted for the operation. A great drawback in most
mills is the want of space, and this will to an extent be
obviated if machines of the plansifter class become general,
as their strongest recommendation is their great capacity.
This system of grading has been more or less left in the
background on account of the competition of engineers,
whereby millers have been tempted to take the lowest
estimate for a complete plant, and the expansion of
machines and also of operations has been cut down to a
very low point.

Next to the substitution of rolls for stones came the law
of grading, and the more carefully it is done the greater
the benefit to the flour, and the quicker it is made. Space
and power do not enter into the question as they did
before; simpler and more efficient appliances have taken
the place of old-fashioned rumbling reels, and although
the principle of good grading has always been acknowledged, reasons, some of which are given, have made it
more or less inexpedient and of a secondary character. If
anything further were required to instil the necessity for

the proper performance of this operation it would be by saying that the more perfect the grading, the quicker the reduction into flour by means of easy purification and smart pressure on the succeeding rolls. The better the classification, and more of it, the better the flour, and that by a shorter subsequent method. The line between pure and impure is thus more sharply drawn ; the material is grouped better at the head of the mill, and this ensures a speedy finish, with little travel, and consequently less break flour. How far this grading can be carried it is not for the writer to say, as it is possible, with accumulated experience, to go farther and farther until every whit of the wheat berry finds its proper resting place under a certain designation.

Wire on scalping sieves does not give the best results, owing to its liability to clog. Silk is too expensive on top covers, as it wears away very quickly, while zinc is found very suitable and very lasting. Wire has what is called a free sifting surface of about 50 per cent., zinc 40 to 45 per cent., and silk 55 to 60 per cent. This means that the perforations in the materials named, or open spaces between the connecting strands of wire and silk, are equal in area to the figures named when compared with the whole surface. In other words, the strands of wire take up as much room as the spaces between them, zinc rather more, and silk rather less. As is, perhaps, already known, sieves are clothed with certain numbers of zinc and silk, and these numbers are the outcome of experience as to what stock from each ought to be available for grading and also for purification. A certain affinity exists between the numbers and the corrugations, speeds, and shearing action on the rolls, and it will be noticed that the finer numbers of each are co-related, as it were, to the previous (i.e., the break) operation. The necessary surface is usually given at two square feet per break for every sack of flour manufactured per hour. Taking, for instance, a 6-sack plant having five

Turner's Vibromotor.

breaks, each sieve will probably be 4 ft. by 3 ft., that is 12 square feet—2 square feet per sack per hour each. These will be employed on four breaks, and the fifth or last will be found delivering its product to a centrifugal, because the amount of middlings left on the flakes of bran are not considered of that quality which will allow of their being mixed with those made before, and also because the product is so much more difficult to scalp and grade that greater force must be exerted to separate the component parts. This exertion a sieve does not possess, and consequently other means are employed to secure the desired results.

It has been shown that in many cases scalping is done by the rotary sieve. There is one which goes by the name of "Vibromotor." It is a flat sieve which receives a rotary motion on a somewhat novel principle. The sieve itself is held suspended by four straps, and in the centre is a bearing for a small spindle carrying an arm with a weight at the end, and having no contrivance to counterbalance. The spindle is rotated by a flexible connection consisting of a rod hooked at both ends ; the weight is not attached to any fixed part, but to that part which has to be moved, and that is the sieve. Thus every action and re-action set up by the rotating weight and the sieve is entirely expended in the operation. It is thus claimed that by the combination of the flexible shaft and the unbalanced weight vibration is not likely to extend to the framework, but is utilised in working the sieve. These machines are made with two sieves, one above the other, and clothed in the manner before described. The makers are Messrs. E. R. and F. Turner, and to judge by the number at work it is a machine giving satisfaction.

Messrs. T. Robinson and Son, of Rochdale, were among the first to adopt the system of rotary motion for scalping purposes, and have steadily maintained its efficiency ever since, and they have developed and improved their appliances until now they are as perfect as can be found. This sieve differs from the majority in that it is not on a level but inclined, and it is contended that the angle is just what is required to make the material fed to it self-propelling, inasmuch as the stock is made to travel over the sieve surface by its own gravity without any other help, without being pushed by the material behind, and this with greater celerity and less abrasion of particles through over-contact with similar particles. Undoubtedly in these days of

Robinson's Rotary Dresser.

trivialities nothing ought to be overlooked to secure the coveted goal in flour, and this is an instance wherein the closest attention must have been given in its conception to bring it to its present perfection. Not only does stock suffer from violent contact with material of its own standard,

but much more so when the impact is against material of
a more or less impure character such as break stock
undoubtedly is. The vertical shaft runs in ball bearings,
by which means friction is lessened and power reduced.
The balancing arrangement is also good, as by the shifting
of a weight it can be adjusted to either the feed or the
throw of the crank ; this attention to detail is continued
even after the feed has left the machine, where in place of
the usual canvas sleeve we find an inner conical device
which guides the stock just operated upon into the delivery
spout, thereby saving friction between the sieves products
and the calico.

In connection with this scalping process it is claimed by
a number of millers that it is still beneficial to have the
blue flour removed after the first break operation. The
crease dirt theory is abandoned so far as earthy matter is
concerned, but we are informed that some kind of mineral
filtrate is deposited there, and in the hardening of the wheat
after maturity the berry closes up tighter and grips this
matter and only the splitting of the wheat will liberate it.
In a proper sense, of course, it is not flour at all, but when
the berry is opened by the action of rollers this matter is
supposed to be liberated, and mixing with the rest licks up
whatever flour may happen to have been loosened in the
process and so discolours it ; hence the idea of a gentle first
break, so as to make no more flour than is inseparable from
splitting the wheat berry open and to provide a fine sieve
or other cover to get this objectionable product away as
soon as possible. With well conditioned wheat, however,
this practice is being dispensed with in favour of a more
severe cracking, whereby 10 to 15 per cent. of middlings
are obtained and the wheat is in a much better condition
for the operation of the second break. In a direct system
where blue flour, crease dirt, or mineral secretion find no
place it will be found usual to place a cover of zinc equalling
No. 20 wire on the top sifting surface of the first four

scalpers, and what material tails over is sent to the successive break roll. The second or under sieve is of silk and is usually about No. 36 of gries gauze quality, and what is refused by the meshes is called large middlings or semolina and from the first three scalpers is sent direct to the purifier, so that in reality the double sieve scalper is also a grader.

The separating of the stock on the fourth machine is rather different owing to the lower quality of its composition, and is much the best treated alone. To mix at this stage by allowing the overtails of the under sieve of this break to join the bulk would spoil all the labour and skill hitherto employed, and so custom points to a finer silk cover, say 60, instead of 36 or 40. The throughs are best sent to be dressed with some secondary purified middlings or rolled cut offs, and treated afterwards as is most expedient. The overtails of the 60 silk may be either purified alone or sent to the centrifugal duster, and afterwards graded with other middlings of about the same milling value. Coming to the last break product no sieve is employed but a centrifugal, for reasons before stated ; No. 4 or 4½ zinc will about suit to get the bran away, and then the throughs are sometimes treated for getting more middlings to purify or else we employ a coarser cut off and roll direct. The intermediate product which tails over this second machine being of the size and value between what is termed finished middlings and bran. If a further development be wanted broad bran will be tailed over a No. 8 or 10 wire, but averages as to quantity cannot be made owing to the various sizes of wheat employed, and that in different proportions by every second master miller one is cognisant of.

Wheats do not always separate with equal freedom, and much difficulty will be experienced when a soft mixture is being milled, and the sieves will require constant watching and cleaning to keep the perforations free and prevent a return of dusty floury stuff to the succeeding break roll.

In a number of mills single sieves only are used, and it then becomes necessary to deal with everything except what has tailed over to the succeeding break in a manner slightly different to what has been already explained. The common way to do this was by passing it to a long reel where the head sheets are clothed with 9, 10, or 11 silk to extract the break flour. Next come No. 5, 3, and 40 g.g. The overtails are the semolina particles corresponding to those of double sieves, *i.e.*, the lower one, and are sent to the purifier, as are also the throughs of the last two sheets, while the product passing through the meshes of the No. 5 is further dusted free from floury material on a centrifugal. This process is also usually followed where reels are employed for the scalping process, the wire numbers on such ranging from 18 to 22 meshes to an inch.

The other machine used at this junction when plants are arranged as above is the inter-elevator, and is, as has been just stated, a sort of compromise between a reel and a centrifugal. Like the reel it has an outer cylinder supported on the central shaft. Attached to the inside of the machine is a set of lifters or elevators (hence its name) and at a certain height these devices continually distribute the feed over the surface of the silk covering. This is a good idea and prevents unnecessary disintegration by the chop falling about too much while being operated upon. This machine also dresses out the flour and sizes or grades the middlings.

The round reel of Messrs. Robinson is also a valuable machine in sorting and dressing the finer products of the scalpers, the chain-like network internally attached being an admirable arrangement to prevent the breaking up of the finer middlings. Gentleness is preached, with extra earnestness, at this stage of roller milling, and it is difficult to estimate the correct idea of some millers in connection

with many of the earlier appliances which were employed
to ensure this effect. But after all the writer is not in
agreement with the mode of separation just touched upon.
In the light of experience, dusting and grading should be
approached from the other end. I submit that single
sieves, chop reels, and inter-elevators do an amount of
positive harm when worked in conjunction, because to avoid
the risk of contamination after each break it is incumbent

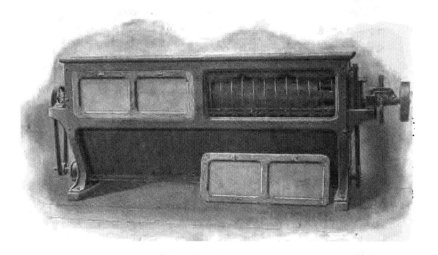

Robinson's Round Reel.

to remove the largest particles as quickly as possible so as
to lessen the amount of friction set up when all are thrown
together again and, as in the case where reels are used, the
break flour already made is increased and by constant contact with branny middlings gets a bad colour which would
not occur if the latter had been removed at an earlier stage,
so that speaking in a broad sense double sieve scalpers are
perhaps the best appliances so far as regards results, and
the saving in power is also considerable when the feed is

distributed in smaller parcels than when it is all collected into one long reel with an intermittent drop and stop motion which characterises the majority of chop reels. For treating the finer material which requires dusting the inter-elevators or round reels with distributing device are very useful machines, being calculated to use just about the necessary amount of friction.

Speaking generally scalpers must be clothed finer as the breaks proceed, and must bear a certain relation to the number of corrugations upon the rolls whose product they are utilised for. They must be clothed finer, as just stated, on account of the size and state of the material which has been reduced. Larger middlings can only be got from larger particles worked upon, and that is at the commencement of the break system, and the middlings will get smaller every time the residue of one machine is again reduced. So that all other things being right for the various breaks the corrugation will determine the middlings and also the scalper covers. The corresponding numbers of wire and zinc run about as follows :—

No. 14 wire is nearly equal to No. 7 zinc.
" 16 " " " 6½ "
" 18 " " " 6 "
" 20 " " " 5½ "
" 24 " " " 5 "
" 28 " " " 4½ "
" 34 " " " 4 "

and these are mainly the numbers employed both here and also in the bran-dusting department. It is important to notice the angle of spouting connected with scalpers, because here we have every variety of product, and it will be as well if a table is inserted giving the number of degrees from the horizontal that each class of stock will travel easily by the aid of gravitation alone. We will assume that no natural or artificial defect is present, and that the wheat before being milled is in what we know as

E

good condition, spouts fairly straight and smooth in the bottom and without any sharp turnings in them.

Flour will gravitate at from 45 to 50 degrees from the horizontal.			
Wheat	,,	25 ,, 30	,,	,,
Coarse Semolina	,,	35 ,, 40	,,	,,
Chop	,,	40 ,, 45	,.	,,
Soft Middlings	,,	45 ,, 48	,,	,,
Light Tailings	,,	38 ,, 44	.,	,,
Bran	,,	36 ,, 40	,,	.,

The points requiring attention in the department of scalping and grading, so as to ensure perfect working being done are :—

1. A constant regular supply of feed spread over the entire surface of each sieve.

2. Regular inspection of stock as it comes to scalpers, and also after the divisions as it leaves.

3. Keep the meshes and perforations open by frequent brushing to avoid dusty overtails.

4. If exhaust be applied to scalpers have enough, but not too much ; avoid condensation and chokes in spouts by frequent inspection.

5. Regulate the throw or area of rotation to get the stock away with as much despatch as its condition allows.

6. Lubricate carefully to avoid waste and injury to break product by over-flowing of oil from any part.

7. If sieves are set at an angle, let it be acute enough to keep from augmenting break flour by too much friction, caused by being kept too long on the sieve.

8. See that the bran does not present a ragged appearance, and that a big percentage of large middlings is present, and so help to fulfil this embodiment of good milling.

9. Examine all parts liable to wear, such as bearings,

hanging straps or rods, and nuts, bolts and thumbscrews, connected therewith, and keep everything in a clean condition.

.0. There should be no vibration in the framework, as this loosens screws and bolts and causes wear of important parts.

.1. Machines should run without the slightest noise, and there should be no escape of dust.

.2. Reels for scalping will run 30 to 35 revolutions per minute, and a sheet of coarse sacking suspended from the top of the framework and allowed to hang on the reel will keep the meshes clean.

.3. To ensure a perfect break product for scalping a certain proportion must exist between the depth and number of grooves on the roll, the particles operated upon, and the number of the wire or zinc sieve which separates those particles.

.4. Angle of spouts to and from scalpers to be acute enough to ensure freedom from chokes, and yet not too straight or perpendicular, else additional break flour will be made.

.5. The vibration mentioned in rule 10 may be caused by wrong adjustment of balance-wheel, uneven bracing of hanging straps or undue wear of the footsteps.

6. Breakages in silk sieves cause some of the stock to travel in a wrong direction and lessen the amount of feed to primary purifiers.

.7. The larger middlings should mainly be of an oval or kidney shape, while small ones are mostly nearly round.

.8. The product to the first scalper will contain about 12 per cent. moisture, decreasing to about 7 per cent. on the fourth sieve. There will be 10 per

E 2

cent. moisture on the feed to the bran centrifugal. This is from well-conditioned wheat and is a fair estimate.

19. A chop reel will take up quite a half-horse power per hour [for every sack of flour manufactured, and should never be overloaded, as the distance from end to end is a great strain on the inside mechanism, viz., ribs and stays.

From the foregoing matter it will be evident that although the operation of scalping is of such a simple nature, there are many little things attached thereto in relation to the material which well repay examination, and from which much may be learnt by constant observation and attention to the proper performance of the duties allotted to them. Among the important things in connection with raw material here is that of proper grading what has been sifted and the dusting of the finest part in readiness for purification. It is recognised as the correct proceeding to secure the best results in the flour. There is no possible arrangement of silk, and no air current that can be utilised to treat effectively widely-varying sizes of middlings at the same time, simply because the natural law of gravitation will not allow of equal distribution of unequal atoms. Therefore, this method of grading must be held to be important so long as milling is pursued on present lines, but into how many grades first middlings must be divided is left to other circumstances than can be laid down in a general treatise. Medium and small mills will usually have three, or may be four, primary sorts, while larger concerns easily run into six or seven.

It is necessary now to take notice of the latest form of scalper and grader, and that is the plansifter. It is not intended at this stage to go into more minute description than is required for the operation. To the mind of the writer, the most important thing about it as a scalper is

that of despatch. The plansifter scalps, sorts, and grades
at one operation, and that, too, with gentleness and
completeness. There is no need of a second machine to
take up the run of the stock, and this does away with a lot
of travel, elevators, and screws, all of which more or less
increase the quantity of the objectionable product known
as break flour. After the breaking up of the wheat
into bran, several sizes of middlings, and what flour
has been made unavoidably, the next principle is to
get them classified without waiting or lying together.
There is always some impurity present at this stage, and the
sooner the goods are put to the next process, by being
separated, the less risk there is of contamination. Here also
the centrifugal scalper scores a point. This chief impurity is
not break flour, but something which the break flour picks
up, and the idea is to make the separations so quickly that the
contact of the component parts of the meal shall be of short
duration and so gentle that further breakage of starch cells
be avoided as far as possible. This the plansifter
undoubtedly does, and taken in this sense alone it fulfils a
very useful and a very necessary function. Just as experience
suggested a reel, a sieve in place of a reel, a double sieve in
place of a single sieve, so now does this machine come
forward with a multiplicity of sieves, which, aided by other
devices for propelling or retarding the feed sent to it,
combines in itself the several attributes which were before
spread over a series of operations and covered considerable
floor space.

Perhaps a short description of one of these machines
will be best, in so far as it affects the scalping and grading
portion of its work. We will take a type. The one under
discussion is that of Dell's, and known as the " Common-
Sense Plansifter," and is constructed like an upright
cylinder in which the sieves are superposed. Both in shape
and action it resembles an ordinary hand sieve. The break
or other stock is fed to it through a central hole, and a

W. R. Dell and Son's "Common-Sense" Plansifter.

distribution is made by a cone delivering it to the outer edge of the first sifting surface, and being delivered at this outer edge what fails to pass through the mesh works its way toward the centre cylinder, and thus drops out for further treatment.   These distributing cones, or trays, are a special feature in placing the largest amount of feed where there is the largest amount of dressing surface, viz., at the circumference, and the residue converges to the centre, and so escapes.   The throw of the crank can be adjusted to suit different wheat mixtures or products. There is an arrangement of loose brushes working gradually round the under side of the silk surface to keep them open, and so assist the machine's capacity, or, at least, not retard it. It is hung in an ingenious manner. The suspension rods can be altered to suit requirements, and altogether it strikes one as being a most useful, ingenious, and effective appliance.

What has just been stated does not, of course, refer to the economy in space—in power, perhaps—and in cost, which a plansifter exercises in this department.   Were any other style of machine paramount, then the same duty would devolve upon me of giving it importance in a like manner.   Flat or inclined sifting surfaces are not yet universal in their application, and while this is so nothing remains for the historian but to state facts, leaving it to fuller and longer practical experience to determine their several merits.   A centrifugal, reel, or single sieve is content to do one thing at a time, a double sieve attempts more, while a plansifter attempts everything, so that, taking this consideration alone, it far supersedes anything yet attempted in a sense.   Whether it has corresponding defects not applicable to its other competitors is just likely. I will say this, however, that whether the dead weight all on one machine for a long period—comparatively—takes up much less power than several lighter loaded ones for a short period is a matter I have not fully determined, but it s a point which will bear investigation by those having the means at hand to do so, because this question of power is a leading recommendation from the manufacturers' standpoint, and ought to be settled.

It is as well to repeat also that besides a revolution in the scalping machinery there has been a revolution in the treatment of the stock. Centrifugals, reels and inter-elevators get rid of the flour first. Sieves of all kinds are built to begin at the other end. There is a good reason for this. All break stock contains middlings, to which are attached bits of bran, and the continual scouring of these middlings through a long reel tends to rub a powder off the bran specks and deposit it with something else, and also to multiply themselves considerably by being broken up.

Sieves, it will be noticed, do the exact opposite to this in getting the larger pieces sifted off, both for their own safety and also for the safety of what remains, the consequence being that there is less breakage and less discolouration, while the whole of it is in much better condition for the next process, for the more granular they are the easier they are to be purified, with the gentle motion they are more sizable and there is less diverseness in shape and condition. Purifiers will best treat material most alike in its composition, so will the smooth rolls to which they are afterwards sent, and the argument can be followed right up to the flour sack.

All makers of machinery add a special appeal in selling or advertising as to the power absorbed when at work, and more particularly does this refer to the department under discussion. There is no doubt that the long chop reel takes a great deal of power, even when run empty, owing to a great part of the machine extending so far away from the seat of transmission, and when a double-chain driven worm is added there is a further disadvantage in this respect. Then again, the swaying and moving mass inside requires the driving belt to be rather tight, and this also adds additional friction, and the stock has also to be continually lifted, making so much more dead weight, while the intermittent tumbling about jars on all the parts. The shorter round reel is a somewhat less aggravation of the points just noted, but the friction of the break stock against the wire covering is much greater than would be the case were zinc substituted, because the feed would not then be carried so high up the sides, less of the midd lings would be broken up, and the travel would be less. On the other hand the advocates of reels argue that some friction or resistance is necessary in order to quickly separate what is being acted upon, and so secure a better provision for the next break. Sieves have none of these faults, if they may be so called, for if set at a suitable angle there is no lifting, there is no propelling, there is not even the nett weight to contend with, because, being set in a slanting position, the probabilities are all in favour of such a quick despatch that some part of the stock is gone as soon as the sieve is reached and the residue does not stay there long enough in the same quantity as by any other means ; so that, summed up, the sieve has everything to recommend it in preference to reels, or other mode or device, so far as is at present known, and the reason for this preference lies in the fact that by it a quick separation is made, thereby incurring less risk of contamination, less power is required, on account of short contact and a better situated dressing surface and less pulverisation of undesirable constituents takes place, and these rules guide us, or ought to guide us, in the duty done and expected at this period of roller milling practice.

Without saying which machine is best it will be enough to remark that where circumstances do not allow of what is perhaps called the best, the best is just what suits the circumstances, and instances are not rare in big merchant mills where all the scalping is done on reels, where all the sizing or grading is done by the same means, and in addition to this where, in fact, the operation altogether is considered of secondary consideration. The wheat is perfectly cleaned and conditioned, the breaking down is by three and not more than four break operations, the scalping is a short handy method of getting overtails for the

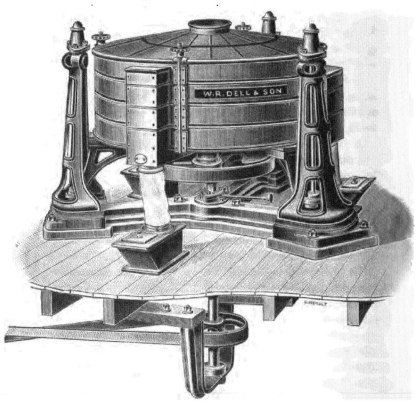

Dell's Plansifter on Floor Level.

next roll, and, in addition, if the finer particles show too much dust they are dusted, if not dusting is dispensed with, and it is sent straight past the purifiers to the smooth rolls. In the case of the larger middlings purification is not thought to be of that supreme importance of a dozen years ago, and if there be too much feed it is sent past that machine entirely and direct to the smooth roll. The argument is, that wheat being perfectly clean does not require the many sub-divisions it otherwise obtained, and that the mill is simply a money-making machine, or ought to be, and, to judge by the results which the writer has cognizance of in a mill doing nearly a hundred sacks per hour, the realisation of the miller's ambition has for a long time been an accomplished fact.

To briefly run through the scalping and grading portion of a mill, as just described, the following notes of interrogation will be helpful in impressing upon the memory the fundamental principles underlying the process :—

1. What is the main idea in scalping or sifting after each break ?

2. What material is employed as a sifting surface, and why, where, and in what form in the various systems enumerated during this chapter ?

3. What are the several revolutions of sieves, reels, chop reels and inter-elevators per minute, and why different ?

4. What principle is involved in sieves different from reels in separating break products ?

5. Why is an inclined sieve considered better than a flat surface, and why is breadth preferred to length ?

6. Describe how vibration is avoided, equalised, or compensated for in two or more different makers' sifters ?

7. How is it possible to get a larger area of rotation if the feed going to the machines requires such to do better work, and what will be the condition of the feed making it advisable to do this ?

8. What becomes of sifted material (throughs) from each sieve, reel, dresser or plansifter, and why is it treated thus before purification?

9. Into how many sizes is it usual or advisable to divide middlings to get the best results on purifiers, and what are the silk numbers employed to accomplish this object, *i.e.*, primary purification only?

10. Give a list of various ways adopted to gain the desired revolutions in sieves, how hung or suspended, rested, or any other device calculated to help in the operation?

11. Learn the table of zinc equivalents, so as to be able to express the numbers without referring to wire covers.

12. What are the relative advantages of silk, wire and zinc in the clothing of scalping and grading machines, and where (usually) are they severally employed, and in what capacity?

13. Describe the treatment given to fourth break roll chop different from the previous three breaks, and give the reason why it is not all dressed and graded together.

14. Give the numbers in zinc with which the first four sieves will be covered, without reference to wire.

15. Name the moving and moveable parts of a vibro-motor reel, inter-elevator and sieve.

16. What special claims does the plansifter possess for doing the duty of scalping and grading as against, or instead of other devices or machines herein named?

17. Which machine in this department is likely to take the most power to drive when at work, both in a direct and also in an indirect manner, *i.e.*, either

by itself alone in doing a large duty or in combination with other machines helping thereto and taken as one against a single machine.

18. Having now reached a very important stage in the manufacture of flour, state precisely what parts of the wheat-berry are finished and sacked away, what parts remain to be treated, and in what condition should these latter now be in to make the next process a thorough success.

CHAPTER VII.

PURIFIERS AND PURIFICATION.

It is, perhaps, one of the drawbacks of milling as now practised that it is a continued series of eliminations. What are denoted as impurities are always present in a more or less degree at whatever stage of manufacture we may chance to look at. Every operation that wheat undergoes during its progress through the mill for part conversion into flour, has for its object the getting away of something from the bulk operated upon. The whole process of wheat cleaning aims at getting rid of extraneous matter foreign to the berry : The break rolls operate towards setting free the kernel and getting away the branny coating; the scalping sieves follow in the same direction by dividing the product thus liberated ; the chop reel, dusting reel, inter-elevator, rotary grader, or plansifter, work according to size in the same direction, and finally the purifier—for which all the machines as yet mentioned operate, and are in fact preliminaries to—has gravitation for its ruling principle. We have already seen how the wheat has been cleaned, and also in breaking down on the rolls how carefulness is practised in the adjusting of each one of them, we have noted the gentle action recommended on the process that follows, whether it be done on reels, inclined rotary sieves, or plane sieves, and the importance which attaches to the grading of the sieves products, all this has been insisted upon with but one object, and that object was the getting of as great a gross percentage of feed for the purifiers in the shape of middlings, and in as large a state as is compatible with the best workmanship. The prime reason for all this carefulness lies in the fact that violent treatment on the partly reduced wheat when as yet the particles are not separated tends to the making of flour, and flour made at those stages is dirty in appearance and bad in quality, therefore, if the quantity of flour made during the somewhat elaborate preparatory period up to the entry of the stock into the purifiers can be limited to from 10 to 15 per cent. at most of the total quantity, the object of it all will have been accomplished. Under any circumstances some flour will be made and leading men have fixed upon the first named amount as a minimum.

Purification then is the finishing stroke of preparation, and once past the purifiers the product is supposed to be ready for immediate conversion into flour. It will perhaps have been noticed in a dry process of cleaning wheat that the valves regulating the amount of suction can be set to a very fine pitch of excellence, wherein the current employed will seize upon the falling grains and weigh them in its balance, first lifting them a little, and anon letting them fall lower down, and seeming to hesitate as to their fitness to join the rushing multitude of their kind until, perhaps, they are finally lifted away as screenings, or may be allowed to go their way to the next machine. It is just this delicate balance of wind power which is depended upon mostly in the process of purification. There is a certain gravitation attached to the multitudinous components of the feed which the wind is called upon to counteract. The amount, of course, depends upon how much these component parts can resist, and, with their own gravity, overcome by falling into the collecting worm, and it is made a more simple performance and a more successful one, because the heaviest particles of wheat semolina or middlings are generally the purest, hence the action of the wind in lifting away or retarding the moment when anything of less weight shall find its way through the meshes of the silk sifting surface, which is always provided as the means of assisting the miller in this very interesting operation.

Speaking in a general sense the middlings which are now supposed to be streaming from the graders are sub-divided into three sections. It has just been stated that specific gravity is the principle worked upon in connection with a

Robinson's Purifier.

generated air current, which has to be overcome before they—the middlings—can be properly classed. The first and best section of middlings are called glutenous—which means they contain a more or less amount of gluten—what gluten is and what is its function will be left to the chapter on wheat—and as a consequence are heavier. The second section comprises what are known as starchy

middlings, and are considerably lighter than the other, and
the third is offal, and to effect a separation of these—to
produce glutenous middlings pure, starchy middlings pure,
and offal without any admixture of either, is the task the
purifier is devoted to, and in what manner it accomplishes
this, taking into consideration the amount of skill required
in manipulation, by that it must be judged.

An explanation of the working mechanism of one of these
modern appliances will alike be instructive both as to its
object, its construction, and the constitution of the material
under treatment. Watching the middlings as they come
into what is termed the feed box of a purifier, it will be seen
that it is delivered regularly and in a constant stream on
to the sieve proper. The eccentric motion of the sieve and
the slight angle at which it is tilted, causes it to travel along
towards the other or tail end. As the feed is thus progressing it is subjected to a current of air continuously,
which is rushing up through the meshes of the sieve and
lifts with it the portion of middlings or offal, which, be it
again noted, are lighter than the other portion which was
heavy enough to resist this wind action, and is either already
through the silk or waiting until a suitable mesh is reached
which is large enough to allow of its so doing.

On the upper side of the sieve of a Koh-i-Nor purifier, and
in close proximity to it, is a curved piece of mechanism
which is called a cowl. The current of air originating at
the top of the machine and caused by the revolutions of the
attached fan, can only be supplied through the meshes of the
sieve, and, consequently, it must perforce overcome the dead
weight resistance offered by the layer of feed before it can
get upwards to its source of generation, and so, to a large
extent, the regulation of the force or speed of this current
of air determines the quality of the work done in so far as
concerns the machine itself. Directly the air has passed
the obstructing layer of middlings—and carrying with it
just so much impure material as the regulating of the

air valve is responsible for—the cowl just mentioned, by
its structural shape, diverts both current and material
brought up into a much larger space, and the consequence
is, that being somewhat freed from the obligation laid
upon it by the just previous rush, the fan loses some of
its hold upon what has been thus far lifted away from
the travelling bulk, and deposits it upon the tray or
platform provided for this purpose, and it is reserved for
other and separate treatment, and what has passed
through the silk openings should, on examination, prove
to be pure middlings ready for immediate conversion into
flour, and will, undoubtedly, be this product if the requisite
care, skill, and attention has been bestowed upon the
operation. It will be understood that once the current is
past this cowl it expands, and it is so close to the sieve's
surface that any material once seized upon and deflected on
to the platforms is finally separated from that of a purer
nature. The platforms vibrate with the sieve, and all that
is deposited thereon gradually slides away. The ascending
air, having now a larger area to fill, is decreased in strength
by expansion, and is more steadily drawn up towards its
source of attraction—the fan—and it meets a sort of
supplementary cowl which, as in the first case, liberates it
to expand again, and in the doing of this it loses its grip
upon a lighter class of impurity, which was not heavy
enough to be deposited on the lower platform, and having
done this duty the air is blown into the open room,
practically free from all contaminating substances.

The sifting surface proper is divided into several sections,
each one of which is capable of being operated upon
independently according to the quality of material under
treatment and the size of the silk meshes, and the latter
are kept free and open by means of travelling brushes
working on the under side. Double worms are a connecting feature to enable the operator to make a sharp
division of stock as required, and by their help in cutting

off any portion of purified middlings and diverting its flow
to any machine to suit any special purpose, are indispensable in a modern machine. Perhaps the greatest
recommendation, however, lies in the even distribution of
anything fed on to it. By this I mean the even distribution, not only when the feed is first spouted to it, but right
down to the last inch of the sieve's surface. It is an
axiom of modern milling that wherever air currents are
employed, to be effective they must be regular, and the
form of this purifier sieve illustrates this in a most perfect
way. If a purifier has the same dressing or sifting surface
at the tail of the machine as at the head, it is quite likely,
especially when working on large semolina, that there will
be bare places when the bulk of the feed has been sifted
through. Wind being very elastic is ever seeking the
easiest way to the fan, and when an opportunity occurs it
will rush through these bare places to the detriment of
good work all over the sieve space. Accordingly as the fan
is speeded it must needs have a certain supply of air to fill
the partial vacuum created by its revolutions, and nothing
short of shutting the wind off altogether from the bare
places will result in having pure middlings from the well-
fed portion of the machine. It may be that the valves are
almost closed on these sections, but that is not a certain
remedy because you must have enough to purify, and that
would mean too much for the other sections and the
general average of work will be more or less irregular and
incomplete. Altogether, this purifier strikes one as being
very novel in principle, simple in construction, and effective
on the work allotted to it. The slight tapering of the sieve
is a strong point when dealing with middlings and semolina. The heavy character of the stock as it travels towards
the tail end needs a strong current, and this it receives
because the more contracted the sifting surface the
heavier the feed, the greater the intensity of the air
suction owing to the lessened space, and these are the very

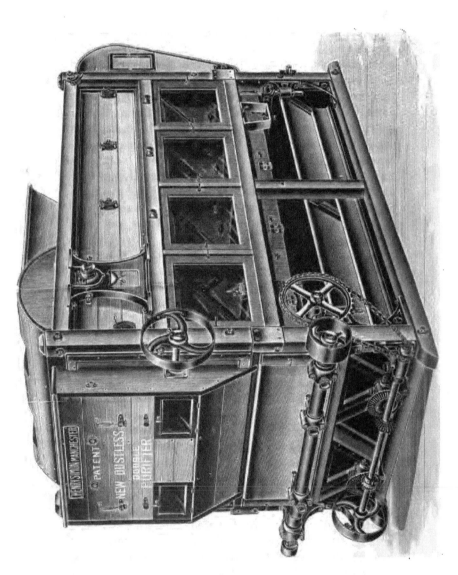

Simon's Purifier.

things wanted to make perfect work; and the contraction also means wider platforms and a greater allowance for expansion, so that, notwithstanding the force of the uprising exhaust at and near the tail end of the sieve, the deposit of light offal and the absence of dust seems to have been hit off in about as neat a manner as one could imagine possible. To allow a purifier a chance to distinguish itself the feed to it should be in good condition, nearly as possible of an uniform size.   This is a very common saying, and it would be a good thing all round if it was as common to act upon it.   A sharp line cannot be drawn between pure and impure middlings without creating a sort of half-and-half product which needs to be taken account of, and the machine which takes this into consideration, as it were, is most likely to be chosen.

One of the latest developments of this system is the "Simon" New Dustless Purifier. It possesses the following special features :—

The tins or channels are brought close to the silk and inclined in plane rearwards at an angle of about 45 degrees, the underline being quite parallel with the sieve..

The construction of the tins is in hinged sections, any one of which can be raised up to allow of easy access to the silk.

The upper part of the machine consists of one single chamber containing a channel dust catcher.   This is composed of grids of tin channels steeply sloped downwards towards the side channels of the moving sieve below, into which the dust is automatically discharged.

In addition to these characteristics, the Simon purifier claims the following :—

The separation of pure from impure material.

The silk cover and the travel of the material are at all times open to inspection.

The aspiration provided for the overtails is easily adjusted and open to observation.

The fan speed is low, and very little air is in consequence required. Should it be necessary, however, to treat very coarse or very dirty semolinas, as strong a current of air as may be desired can be used without blowing dust.

The construction and finish of the machine is in good style. It is provided with balance wheels and lubricators and means for sampling and examining the products.

Special sizes can be built to meet special cases.

The double machine is really two single and independent purifiers built in one frame for convenience and economy of space; each sieve is distinct and separate from the other, and so arranged that the throw of one balances the throw of the other.

This purifier is also fitted with a self-tightening silk cover.

This is an ingenious contrivance which facilitates and simplifies the putting on and tightening of the silk cover evenly across the whole surface of a sieve.

All purifier, scalper, and tailings sieves are supplied with this arrangement.

The even and open flow of the material on a purifier sieve is absolutely essential to good work, and this the invention secures.

A still further point lies in the fact that the silk is evenly strained over the whole sieve surface.

The bearings are of a new patent pattern and are entirely self-oiling. In fact, all Simon's machines are fitted with continuous automatic oiling devices.

It is quite possible to have good results from purifiers turning off nothing but high-class middlings and low-class offals; but with the catching off of an intermediate product a higher standard is reached in middlings purification.

The feed should be kept within a moderate range of sizes for one machine, the extraction or deflection will be 10 to 15 per cent. for re-purification or other separate treatment,

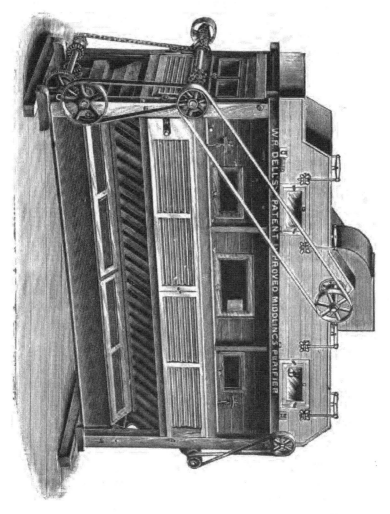

Dell's Purifier.

the tailings will average, perhaps, 10 per cent. of the total quantity fed on, so that in reality 70 to 75 per cent. should have passed through the meshes of the silk fit for reduction or smooth rolls. The air valves must be regulated to avoid loss or inconvenience by over extraction; at the same time nothing should be allowed to slip through which will tend to lower the quality of the work done and to vitiate the stock going to any following machine. It may be here said that all purifiers aim at accomplishing these points in some part of their construction. There is a certain amount of pure stock and the task is to get it away from the bulk, and whether it be deflected or cut off in any other manner does not matter for book purposes. Suffice to say that the more perfect the separation the better is the work considered to be done, and the class of machine doing it most economically, bearing in mind local conditions and surroundings, is the one to keep its hold upon the individual user. It is important that the feed going to purifiers should be free from flour—that is, should have been thoroughly sifted and dusted, because in the event of its not being so the same air current which is engaged in lifting out the impurities will also carry this dust upward and deposit some of it on the platforms, and cause much more trouble to separate eventually than if it had been done at the proper stage, and what is not so deposited will perhaps find its way through the sieve with the best middlings, and so neutralise all the good of purification. Remembering that this dust can be traced back and identified as break flour, the connection between them will be readily grasped by all intelligent readers; in fact, there is scarcely anything that will more amply repay any trouble taken than the fact of having the feed to purifiers thoroughly freed from dust adhering middlings.

Another machine of this class we may pause to examine has for its leading feature a vibrating tray of nozzles covering the whole area of the sieve surface, and a large

number of holes are made in the tray close to each other.
Into each of these holes is fitted a nozzle, on the top side.
They are hollow, and wider at the base than the top, for

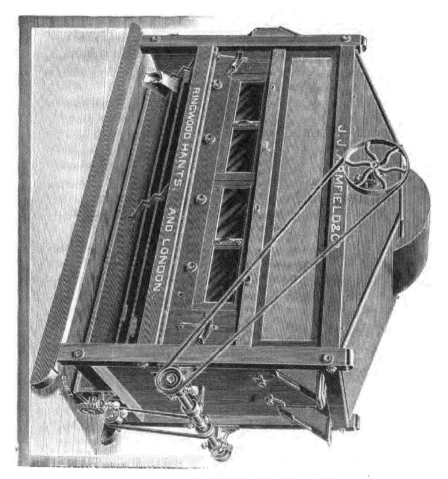

Armfield's Purifier.—*J. J. Armfield and Co., Ringwood, Hants.*

the purpose of causing the air rising through their interior
to become accelerated, and immediately it reaches the

small orifice at the top it expands, and thereby the
impurities carried up through the nozzles are deposited on
the surface of the tray, from whence they are removed by
the vibratory motion, which is common to all machines of
this class. In this purifier every condition of working and
detail of the machine has been very carefully considered for
the purpose of utilising the exhaust air currents in the
interior to the utmost advantage, and to render the purifier
simple and adjustable to all variations of stock, either
caused by wheat mixture alterations or atmospheric
changes. Each section is provided with duplicate sieves,
which are inter-changeable, and there is practically no
limit to the experiments which it is possible to try in order
to hit off the best of anything fed on to it. The superstructure is also on the same lines as the tray of nozzles,
and provides for the deposition of the lighter fluffy tissue
which escapes the action of the quicker wind motion during
the first separation from the heaviest material.

A common feature of all is to exclude all inlets of air,
save what comes through the silk meshes, so that the
efficiency may be the more attained, and to this end they
are fitted with some non-textile padding wherever there is a
constant necessity to remove or take down any door or
window connected with them.

It is also claimed that, owing to the construction of the
nozzles, the current is evenly distributed over the whole
section under observation, thereby rendering the purification
regular at each portion of the sieve area. The tray of
nozzles is set very close to the sieve, rendering it an
impossibility for anything in the shape of light starchy
middlings or branny impurities to fall back after being once
lifted away. The distance is so short and the air becomes
so much more intensified in velocity that they are at once
whipped through the nozzle and deposited on the tray to be
removed as stated above by the vibratory action of the
prime mover, viz., the eccentric.

Practical millers are well aware that a considerable range may be obtained in the capacities of purifiers by a judicious arrangement of the silk numbers on the sieves, according to the quantity and quality of the feed going to them. As a general rule the makers advise longer machines for the finer qualities, and short ones for the coarser middlings, and this is in effect bearing out the statement made or implied that a reduced feed requires a reduced area to do its work, as the line of demarkation is more sharply drawn here than is the case with fine middlings or dunst. The purifier under notice has, I think, had an extensive sale, and its best sphere is undoubtedly upon the softer and more friable portion of mill stock where gentle action connected with sufficient sieve space best separates this somewhat obstinate product. So far it has been assumed that everything is in first-class order, and that no difficulty has presented itself to spoil the ideal purification made famous by all writers on this subject. It is a well-known fact that a great deal of variation exists in so-called similar wheats, and the action of purifiers will vary, and so will the results in exact proportion to the shortcomings of the raw material, and if the wheat cleaning has not been done in a careful manner— that is, if the wheat has not been properly purified in its whole state it will be impossible to attain the highest results at the present stage. I have already stated previously that middlings must be made as large as is consistent with the system, and as many of them as can be had, but dull or badly set rolls will spoil all this, and also according as the dusting and sizing, or grading portion of the plant gives out its work, so will the purifier follow suit — good, bad or indifferent. The purifier is built upon a certain plan to deal in a certain way with a certain kind of material; let the material be not what the purifier is built to deal with and all calculations of efficiency are scattered to the winds. No doubt

the imperfect cleaning of the wheat is answerable for the
largest amount of hindrance to the highest results to be

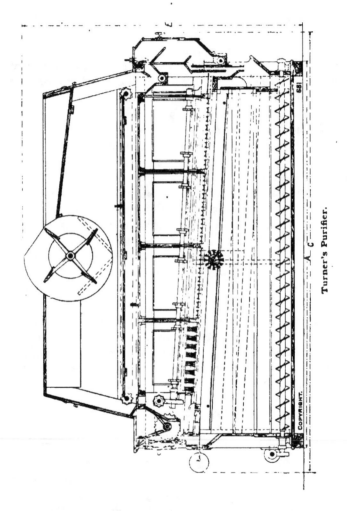

obtained; in fact, the state of the wheat feeding the break
rolls is accountable, in a large degree, for purification even

before it reaches the purifier, for should it not be perfect and contain even only a small amount of dirt particles, these latter, by specific gravity, are sure to be found damaging the best middlings. The minute atoms of earthy matter travel all the way, and will surely deposit themselves on the best middlings, and from here to the flour sack is only a short certain cut. Nothing can atone for dirt passing the rolls, and nothing can stop it from arriving at the above destination. The omitting to keep break roll corrugations sharp also greatly hampers correct purification, dull corrugations lessening both the quantity and quality of the middlings made. When the corrugations are fairly sharp on a proper differential the middlings are clear and well shaped, plentiful, and not so much burdened with extraneous matter such as bran snips, either adhering or loose. On the opposite hand, let them get too dull, and it will be found that the product under discussion is lacking all round; they are broken up too much owing to the rolls having to be set up closer, they have a dull steely cast about them, and when they are dusted and graded the operation is inefficient and unsuitable for the sieve covers over which they have to travel. The weather also here as elsewhere plays some part in the general chaos, for if it be damp and foggy the difficulties are evident all through the mill. Just here it will be found that the dusting and grading is a difficult thing, as the particles cling together more or less tenaciously, the purifiers cannot separate so well, and there will not be the same amount deposited in the trough conveyor, and the overtails will require more attention because of the greater quantity of them. Should, however, the other extreme exist in weather conditions by being particularly dry, hot, or bracing, nearly all these latter obstacles to good workmanship at this stage of flour milling will disappear, and the task be a comparatively easy one in this respect. The work on a purifier is also affected by varia-

tion in the speed of the mill. This will be readily understood when it is remembered that one revolution per minute
more or less of the engine crank shaft means a difference of,
perhaps, a dozen on a purifier. Slipping belts are also
every day troubles.

Purifiers should have enough but not too much feed.
Middlings from moderately hard wheats present fewer
difficulties than those from soft wheats, as purifiers can
carry a bigger load from the former, and the consequences
from overfeeding also vary with the size of the material.
In the case of coarse middlings the overtails will be too
rich, and with fine middlings the throughs will lack
brightness, and will not be so clear looking as they should be.
If these machines are underfed some portions of the sieve
will be uncovered, and the air which should pass through the
middlings will rush up where there is the least resistance,
and the result will be seen in the stock from some parts of
the machine not being purified at all. If the sieve covers
are not put on tight and straight they will sag and impede
the travel, frames must be set even, suspenders placed in
proper opposition to the eccentric stroke, air outlet not
cramped up to cause back pressure, and the silk surface
must be kept thoroughly clean by the travelling brush.
By each section being self-contained and the air current
under control, a very gentle action may be used at the head
of the machine where the fine middlings are being treated,
and the strength of the current gradually increased in each
succeeding section to the degree required, as the material
changes its character continually during its short travel
over the sieve.

It is always best that purifier feed should be spouted
direct to the machine, as worming it breaks up some of the
stuff and makes dust. This applies specially to very fine
middlings, as, next to tailings, this product is the most
difficult to purify, and requires quite a large surface area in
comparison to the larger and heavier class, and wider

machines suit it best because of the shortened distance it
has to travel, and, therefore, less risk of being damaged by
excessive friction.

All sieve purifiers have one thing in common, and that is
to concentrate the energy within as small a compass as
possible, and so allow of the greatest possible amount of
intervening space for the purposes of expansion, and the

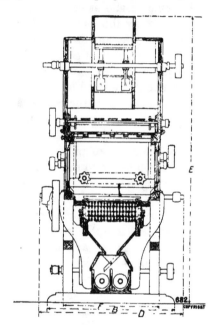

Turner's Purifier.

only real difference in the construction of the principal
machines in the market is in the manner of compassing
these ends.   As a further instance of this may be
mentioned the well-known " Premier " purifier invented
by Mr. J. M. Emerson.   There are no nozzles or
platforms and no tapering of the sieve in the sense
we have as yet noticed ; but still a narrowing of the

available dressing surface does take place. Taking an
ordinary size, for example, we shall find it divided up into
three parallel sections of an increasing width to each
section as shown in the illustration, and by this means the
sieving surface is contracted from head to tail, so that the
tail-sheets get the smallest silk surface and the head-
sheets the largest. It does not arrive at this result by
anything like the same means as others, but the principle
must be admitted, and by this is intended to be conveyed
that a greater area for dressing is provided at the head,
where the bulk of the feed is, than at the tail, where three-
fourths have been disposed of, and, inversely, less expansion
space is required where it is wanted for other purposes, and
more expansion is allowed for at the tail, where a stronger
current obtains, and this is done solely by internal arrangement in all cases. The feed on this machine runs in
channels, and these channels are provided with adjustable
lips, or wings, which can be set to increase or diminish the
intensity of the uprising air current on each section as may
be advisable according to either the feed or the quality of
finished product required. Besides altering the current
by the adjusting of the movable wings, two fan speeds
are provided for, and there are also auxiliary slides
in the expansion chamber. Cross spouts are placed
underneath the sieve to separate the different grades
of offal that settle in the channels between the
head and the tail sheets. The sieves are easily detached
and replaced to meet a quick change of dressing numbers.
There is no doubt that a great amount of ingenuity has
been expended on this appliance, and that it is acknowledged is proved by the large number to be found at work
in all parts of the kingdom.

The guiding elements in building machines called
purifiers are these :—1st, to get the impurities away with
the smallest expenditure of wind and space, as witness the
closeness of the main device to the floating feed in the way

of deflectors, nozzles, perforations and adjustable wings; and 2nd, to allow the greatest possible open space afterwards, so that the lifted impurities may at once drop down

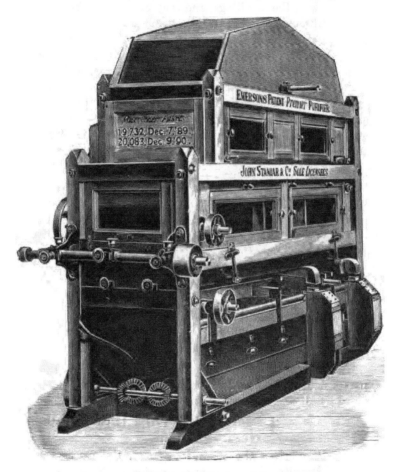

Emerson's Purifier.—John Staniar and Co., Manchester.

again on a separate resting place, and not be carried up to the fan and cause serious inconvenience and loss. Before

F

the introduction of these air compressors fans had to run
500, 600 and even 700 revolutions per minute in order to
seize upon the impure material and prevent its sifting
through with the middlings, and they had also to be provided with dust chambers to collect and settle what it was
an impossibility to deal with inside any purifier ; but now
with larger fans run at a quarter the speed and with these
clever internal arrangements, we can contract, expand and
deflect two or three times and get rid of the air used in the
operation in a comparatively pure state without resort to
settling chambers or dust rooms.   This in itself is a vast
stride in the direction of cleanliness, saving in space, less
risk of dust explosion and better results on the material
operated upon.

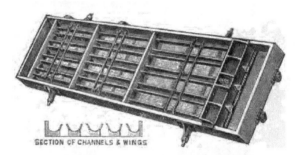

SECTION OF CHANNELS & WINGS

Upper side of Sieve of Emerson's Purifier.

The following are a few suggested rules bearing on this
part of flour milling, and would be useful at times if remembered by all students reading these lines :—

1. The stocks must be well dusted to ensure satisfactory results from purifiers.

2. The better the stock is graded, the better the
   results will be ; a number of sizes of middlings
   should not go to the same machine.

3. The feed must be kept regular and constantly
   supplied.

4. Speed must be kept up, so that the vibrations arising from the eccentric are even and regular.

5. Brushes to be kept well up to the silk and, in the case of softer or lighter stock, supplemented with hand brushing.

6. Suspenders or hangers properly controlled so that the feed is made to travel the right speed down the sieve to ensure just the amount of work being done which the nature of the stuff requires.

7. Large steady-running fan, powerful enough to do what is wanted without having to open the regulating valves to their fullest extent.

8. The machine should be perfectly air-tight—no air allowed to get to the fan except through the meshes of the silk.

9. There should be ample room in the diaphragm of the machine for expansion, and to compass this the contracting space and period will be found to be of the smallest.

10. The rooms where purifiers are at work should be well ventilated and a constant supply of fresh air available, instead of the continued return of the same for successive duty, because in the case of the latter condition of things the air gets stale and moist, and the temperature is increased beyond its natural surroundings, which leads to further condensation and consequent mischief.

11. If the above condition of things becomes aggravated a ventilating fan ought to be put up to create a free circulation of pure air.

12. The divisions or sections of each machine should be perfectly independent of each other; there should be a separate valve for each section, so as to regulate to a nicety the amount of air requisite for that particular locality and condition of feed.

F 2

13. Cut-off worms should be on all purifiers to allow of any part of the throughs or middlings being diverted to any other machine that may be necessary or convenient, apart from the bulk of the product.

14. Plenty of aspiration on the overtails, to correct or further help the efficiency of the general machine.

15. The finer the stock, the greater should be the slope of the sieve by the aid of the suspenders and thumbscrews.

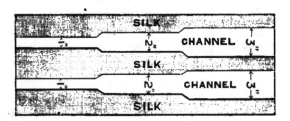

The Silk Surface and Channels in Emerson's Purifier.

16. Judicious arrangement of covers to enable the greatest quantity of work to be accomplished with the least travel of feed.

17. If more impurities are noted in one place than in another, the feed is thinner there and requires a readjustment of the sieve.

18. Brushes and chains to be kept fairly tight by means of the moveable bearings and set screws.

19. The deposit platforms, trays, channels, canals, or other contrivances should be kept free from accumulation of light fluffy material, and breakages in sieve covers should be at once detected. Do not patch too much, rather renew, as patches interfere with the efficiency of the machine.

20. Eccentric boxes to be kept right by the set block
    and hand-wheel; set up without knocking, and
    allow just that amount of play which ensures cool,
    .   quiet running.

A machine meriting description is that of Messrs.
Turner, which has several distinctive features, among them
being an arrangement whereby the trays or troughs are
arranged laterally, that is, across the sieve, and are easily
removed in sections, and the offal which the fan causes to be
deposited in these troughs is removed by a revolving brush
to side canals, which then convey them away to a separate
spout. The brushes can be stopped at any time, and the
offal allowed to accumulate for such time as enough is
deposited to enable the operator to examine it, and also
this test is acted upon to see if the deposit is the same
quite across the sieve in quantity and quality, and also to
gauge the force of the air current, as, for instance, should
the feed be heavier at one side than the other, the force of
the air-current is diminished, and less material will be
lifted than on the opposite side, and, as a rule, these
cross-troughs are set far enough apart to give the man in
charge an opportunity of inspecting the travel of the feed
between them. Like others before mentioned, it is divided
into separate compartments or sections, and can be treated
to a different dose of wind to suit what is at any time
floating over that particular section. The superstructure
also is built upon the same lines as the majority of dustless
purifiers in regard to principle, being designed to intercept
the bulk of light dusty tissue and deposit it within its own
compass. Aspiration of the tailings as they leave the
machine is also provided for in a somewhat novel manner,
and altogether the build and finish of the purifier reflects
great credit upon its inventors. The fan is large, and
makes but few revolutions in comparison to the types in
general use before engineers fully grasped the laws relating
to the displacement of air.

Perhaps some readers will wonder how it is ascertained
which numbers of silk are most suitable for these machines.
Of course, it is easy to see that they must bear some relation to those on the previous ones, viz., graders or dressers.
There is not a great diversity of opinion on this subject;
but of the number of young millers reading about this,
and being suddenly placed in a position where some alteration had to be made in either or both, not a few would be
found stranded and unable to clearly follow the right
course.   In order to put an imaginary case we will take
one or two instances tending, perhaps, to impress this
point on the memory, because the writer painfully
remembers many personal blunders before being thoroughly
able to grasp all the details attending the operation of
clothing purifiers to give the best results. Primary
middlings—middlings which have not passed through
smooth rolls—may be divided into several classes, and each
class is gauged by size alone, and taking the largest we
shall find that, as a rule, it is composed of mixed pieces of
stock which will not pass through the meshes of the under
sieve of scalpers, and the latter may be clothed with gries
gauze No. 40.

These large middlings go straight to a purifier, and
the number they will *not* pass through on the previous
machine is the guide in beginning the work on purifier
covers. Some millers prefer a fine sheet at the head
to sift out anything in the shape of flour which
has been made during the travel of the stock
after leaving the grader.   This is a safe plan, and ensures
a better result than could otherwise be obtained, but it is
at the cost of sacrificing a portion of each purifier to do
duty foreign to its generally recognised nature.   Coming to
the sifting surface proper, a 32 silk will best suit all round
purposes, then 28 and 24, and perhaps some will say 18 at
the tail; but if this is done, then the tail sheet must be
rolled alone, as it is far from being free from branny

particles. As a matter for remembrance it may be put
thus : A semolina purifier begins where the scalper leaves
off, and graduates some 20 meshes from head to tail, each
sheet being a little coarser in mesh as the end is approached.
This is found to answer very well in practice, as it allows
the best round heavy middlings to get away first, while the
4, 5, or even 6 sided pieces with bran attached do not find
an exit until the last sheet is reached, and if this is 24 then
the chances are that they will all tail over to a scratch roll.

We have to select the next size of middlings from the
bulk of the chop as it issues from reel or other graders, and
the number of the silk or wire it will *not* go through is 56.
Following this up to the purifier we should see that very
good work could be accomplished by having the head sheet
clothed 70, and graduating down to about 50 for the tail
end, and with a proper amount of feed, and that in a dustless condition, pure middlings can be made all the way.
And here again, it will be noted, a difference of 20 meshes
is allowed in the length of the purifier, and this will practically exhaust as many sizes of pure middlings as are present
in the product.   The third example of primary middlings
are the throughs of the 56 above mentioned which have
just passed over a No. 6 silk ; and on the purifiers working
this, and also on machines of this class working fine stock,
a bigger range of silk numbers is generally allowed for.
Thus the example under notice will perhaps have 80 meshes
to the inch to begin with, and dwindle down to 50
at the outlet; and all these are not patent middlings,
because the glutinous portion, or a large part of the
glutinous portion, of the middlings has already been dealt
with on the previous machines ; consequently these are
more starchy than glutinous, and except, perhaps, on the
first half of the surface of the sieve they will be sent to join
the straight-run grade.   What is left of the whole of the
break chop is more or less dusty, and opinions vary as to
whether it pays better to send the throughs of the No. 6

silk to a centrifugal for dusting purposes, or send direct to
smooth roll and mix off the resulting flour in some
ordinary grade.   I should deprecate the above facts and
figures being taken in a literal sense, as wheat mixtures
vary so much in themselves, in their composition, their
condition and the class of flour trade cultivated, that
perfect accuracy is not to be thought of save in, perhaps,
some isolated instance.   What is intended is that the
miller may take them as a general guide, and within a
certain limit, as rules laid down to point out the way of
understanding the principles underlying the operation of
clothing purifiers in order to obtain results varying to an
almost limitless extent, according to local conditions and
requirements.

In secondary purification—that is, purification of stock
which has been rolled and dressed once or more—the
numbers of silk covers usually range from 70 to 44, or
even to 40, and the overtails of such are either sent direct
to the offal sack, or, if wished, to a separate tailings roll.
The single consummation is separation of light from heavy,
large from small, brown from white, and finished from
unfinished ; and this simple definition will perhaps be
excused in a book designedly written for the younger generation of those who have adopted flour milling for their
life's vocation, and who, by means of books and study, are
determined to know as much as possible how the science of
it is practised.   In all the types as yet mentioned, single
machines only have been described, but most makers provide a double machine in one frame, divided one from the
other by a board] down the centre, and the eccentrics are
placed or run opposite each other to counteract and balance
the vibratory motion caused thereby.   This arrangement
answers best if two fans are provided so that the work and
action of each is entirely independent, but in the double
machine of an earlier date one fan was expected to do duty
for both sieves, but when the material is of diverse quality

—as is often the case—the result is not so pleasing, and a good many millers who thoroughly grasp the supreme importance of the operations of purification, rather than be put off with a half-and-half sort of system, have gone to further expense in having a large number of single machines to enable them to do the work more thoroughly and with as little regard as possible to everything but what was on the one sieve at the time of inspection.

In many purifiers the feed is delivered perpendicularly into the hopper, and the feed gate is usually weighted and so adjusted that when it is sufficient to spread right across, the weight of it shall just overcome the resistance noted, and so secure a thin, regular flow all the time. Should the feed become uneven, and thicker at one end than the other, a bit of sandpaper rubbed gently on the inside of the feed gate will usually do all that is required. If there is any vibration in the eccentric shaft when at work, alter the position of the weight in the balance pulley until the shaft runs quite steady and quiet. As a rule, also, the brass bearings in this shaft are interchangeable, so that the taking up of the wear can be accomplished for a comparatively long period if given the proper amount of attention in oiling, cleaning, and general oversight. In working purifiers it is a convenient rule to keep the clothing as fine as possible, run the fan as slow as is consistent with good results, and the connection of the discharging or other spouts made with flexible sleeves should be kept perfectly air-tight. Ordinarily the makers of purifiers provide different sizes of pulleys in extra quantity to what are in use, and it is very interesting to experiment with these on the many qualities of feed with which milling students are by this time somewhat familiar, and a great deal can be learnt in this way, given an intelligent insight into the possibilities attainable by this means, and master millers would do well to indulge the enterprising apprentice or journeyman in this respect, because whatever may be gained

is not solely for the man's edification, but must certainly
enhance the, perhaps, already good reputation of the mill's
turnout.

There is probably no part of the science of flour milling
which has thus far had the care lavished upon it that
purifiers and purification has. Latterly it must be admitted that the wheat in its raw state bids fair both to out-
strip everything in this respect and also to, perhaps, lessen
the importance of the operation we are now discussing;
and yet, in face of that, improvements are still going on in
almost all the purifiers we have glanced at, and what may
be considered the chief characteristics to-day may be
superseded to-morrow. For instance, millers and engineers
have thus far pinned their faith successfully to sieves
enclosed in expansion chambers, and allowing for the full
display of a natural law, and by the aid of deflectors, cowls,
baffle boards, and the like, compassed the desired end.

Re-purification is a secondary operation performed upon
the material which the first purifier omitted to act upon—not
necessarily from any fault, but because it is mixed with
every kind of feed impurities, which either adhere to or are
of nearly the same specific gravity as those middlings of
pure build and not troubled with adhering substance. As
an instance of this may be mentioned the throughs of the
latter sheets. If examined minutely they will show up a
brownish mixture in comparison to what has sifted through
before, and this will be more noticeable as the stock gets
farther down the mill. In the better runs of purifier feed
it is best they should be gently rolled and dressed through
a centrifugal or other dresser, and a fine sheet provided as
a cut-off on the latter, say No. 5 or 6 silk. This is fit for
reducing again direct, but the coarser particles which tail
over this number are usually sent to a purifier to re-purify,
so as to get a further separation in order that the smaller
sized starchy middlings may be obtained relatively pure
and the tailings treated according to the run of the mill's

product. One point respecting purification has never been unanimously arrived at, and that is in regard to the treatment of the last break roll product. A few years ago it was the custom to treat this material with contempt, and any slipshod method of getting it finished was considered good enough for it. Formerly five, six and seven breaks were more common than now, and certainly in the latter case it would be a waste of energy and machinery to attempt to improve anything proceeding therefrom by purification, and herein millers were quite right. Gradually the latter breaks were discarded until five was the standard number, and then the other extreme was put to work, that of elaborate purification. Now, however, I should say the standard number is four breaks, and I am aware of some mammoth mills using but three, so that circumstances have completely altered in regard to the quality of the latter break roll products, and what was right a short time ago does not commend itself at the present time. To put it as plainly as possible, I should say that all large mills having five breaks will find it advantageous to purify the throughs of No. 40 wire on the bran reel, and those having but four breaks, more so, on any plant over 10 sacks an hour capacity. For mills of smaller output I am doubtful whether the trouble and expense incurred will be recouped, and for very small mills it is not to be thought of except in special circumstances. If a mill's break system is limited to four operations, then it is advisable on all but the smallest to treat the last break product on purifiers, and if there are but three breaks it will pay everyone to do this; and apart from the latter deduction it is advisable on account of the amount of usable stock—it can hardly be called middlings—that is adhering to the bran flakes, which, if simply rolled and dressed after sifting out these bran flakes, will be a dark brown colour in flour, whereas separated according to the suggestion just given, the difference between the two flours is a great gain on the

colour, and more than repays the trouble taken in limiting
the low grade to its proper constituents. Purification, then,
is only limited by the fact of its being beneficial so far as it
can be carried without rendering unsaleable any residue
not so treated, and without costing more to perform than
the mill's usual quality output would seem to warrant in
every-day practice.

With regard to the number of purifiers necessary to any
given plant it may be stated as a general consensus of
expert opinion that about 900 sq. ins. per sack per hour
capacity is taken up for primary purposes before the
middlings have been to smooth rolls; as, for instance,
taking a six-sack plant, it will require three purifiers
measuring 60 inches by 30 inches to treat first hand
middlings emanating from the scalper stock.

Thus, $\dfrac{3 \times 60 \times 30}{6}$ is equal to 900 square inches per sack.

Concerning the amount of surface for secondary purification, the facts are impossible of tabulation, so great are the
differences in which engineers indulge. The district, class of
wheat used, and trade cultivated whether large or small installation, and the generally defined ideas of millers themselves
relating to cost and power—these and many other things
not to be thought of at random all influence the arrangement in respect to after purification, and while giving it as
a general axiom that the later should at least be equal to
the former, the writer is totally unable to gauge an
approximate number of inches by which to govern the
general run. Small firms are contented with very little
purification, on account of the probability of increasing the
classes of middlings beyond the number of reduction rolls,
while on the other hand the newer and larger mills arrange
for purifying almost everything, regardless of the length of
system. Others, again, after a fine erection of machinery,
turn their hand to cutting and chopping up the flow-sheet
to suit their various ideas, but this is not usually the case

with the best men; they are very careful in this respect to
look not only at the direct result of any alteration, but to
weigh the consequences of any change.

There is, however, this distinction to be made—that
whereas the smaller class of inland millers use perhaps a
great proportion of English wheat, and this is apt to part
with its flour quickly, so far as purifiers go, and being of
itself clean and somewhat tough of nature, primary purification is sometimes the only essential. Far different is it
when a large port or town mill elects to use all foreign
varieties of wheat, which, as a rule, are hard in structure,
and turn out a larger percentage of middlings for the
purifiers and also a larger proportion of branny impurities
with them, and this necessitates that after a first rolling
and dressing it is incumbent to re-purify, perhaps twice
repeated, before the final touch is given by the smooth rolls.
There is not much difference in the first purification
between any two kinds of wheat or wheat mixtures—it is
after the first rolling or reducing that they show the
greatest variation in quantity—and so to say how much
surface is required for purification, as a rule, and
make it a hard and fast line, is beyond the scope of
any ordinary man. It will be understood, I hope,
that while in the one case middlings are soft and
in, perhaps, smaller quantity, on this account they require
more surface than in the case of hard stock, which
separates freely and gets rid of a larger quantity on the
same surface area, so that practically the first middlings
are equal in quantity so far as this surface is concerned,
but once past these purifiers there is not usually that
necessity for re-purifying as exists in the matter of hard
wheat, because the line of pure and impure is more sharply
drawn, and the former are practically more free from bran
chips, and so can go to a further reduction with less
danger than would be the case were their nature the
opposite.

To go over the ground of purifiers and purification in the form of giving a few questions which involve the main principles treated therein may help to engraft into the reader's mind some of the essentials connected with the operation, and may also be of benefit for future reference :—

1. Purification : What is it, and for what purpose is it considered necessary in modern flour milling?

2. What is the best state for the feed to be in so as to get the highest results ?

3. What machine or machines are best adapted to preparing coarse, medium and fine middlings for purifiers?   Give the number of the silk mesh through or over which the various first grades given above separate, and why do they not go together to one machine ?

4. Explain three different ways of compressing and expanding the air drawn through the sieve, and state the law governing all such operations ?

5. What are cut-offs? What purpose do they serve ? State any particular difference in their composition from that not so dealt with.

6. Point out some of the difficulties of purification when stock is not dusted properly, not well graded, and when in bad condition owing to the damp condition of the wheat.

7. Give the remedy for the state of things mentioned in question 6, and if more than one specify in detail.

8. Mention the chief moving parts of a sieve purifier, usual sizes of driving and driven pulleys and fan, and speeds of same.

9. Select one of the machines described in this department and specify the stationary parts inside the framework, for what purpose they are there, and the manner of doing their duty.

10. Explain the system of clothing and numbering the silk covers on purifiers, and the general range of sizes of mesh on (a) coarse middlings, (b) medium ditto, and (c) fine, and also what is the general rule in arranging same.

11. Give a short rendering of the meaning of the terms deflected, diverted, contraction, expansion, and deposition in regard to the treatment of air-currents as employed on purifiers.

12. What advantages are said to be gained by a slow-running fan as against one making three times the revolutions, seeing that in both instances a certain volume of air must be generated to do the work properly? What principle is involved therein relating to a natural law?

13. What is expected of automatic brushes? How are they fixed and driven? Give a list of connecting parts.

14. Why is aspiration made stronger as the feed gets less, and what is the object of again aspirating at the tail end of purifiers after it has passed the whole length of the machine?

15. What is preliminary, primary, and secondary purification, and where does it take place? What are the special objects sought at each stage mentioned?

16. How is air conducted to the fan, and what is the sole reason for its being so treated? State what would be the result if open spaces were allowed in doors, windows, or discharge spouts.

17. Purifier covers are arranged in sections; what is the chief reason for this, and what would be the result were they left open the full length of the sieve?

18. How is it usual to take up wear and tear, avoid vibration, and accelerate or retard the flow of the sieve stock under operation?

19. What class of middlings require most and least wind respectively?

20. Is it advisable to run delivery worms of purifiers slow or fast, and if either, why so? and why is long distance worming considered detrimental at this stage of milling?

21. Taking the whole operation of purifying in its entirety, what should be the prime result in the flour sack as against a system of milling which allows only for primary or partial employment of these machines?

CHAPTER VIII.

REDUCTION.

According to what has already been written we have arrived at a very interesting stage of flour manufacture, and, taking note of all the preparations which have gone before, it should not be very difficult to understand what is now required. The author does not by any means suggest that the close following of written matter by milling students will enable them to become first-class millers at a bound. No one can do that, because it is only possible to rise to the height of knowledge after years of steady practical labour. What is intended to be taught is this, that, seeing that a separate system, as it were, is involved in every operation of flour making, certain rules must be observed every time the product is touched. The explanation of the why and the wherefore of these rules is a starting point from which may spring a deeper interest in the work itself on account of the fact that, being aware to an extent as to what ought to be the final result in each department, the practical handling of the machinery will be greatly facilitated by the student being able to see whether or not he is getting the results he has read about as he goes along, and, if not, then his book learning is of benefit in telling him he is wrong. Book learning alone will never make a miller, but it will greatly help in this direction by pointing out what ought to accrue from the practice of milling. Some definite object is sought after

at every stage, and the simple fact of knowing when in
practice that object is attained is, I take it, the sole aim of
all writers, and having this technical knowledge at his
finger ends the miller's progress in the practical direction
should be all the more rapid in saving him the extra trouble
of finding out for himself what everything means and the
reason of its adoption, and his learning will be more solid
seeing that it is founded upon the experience of those who
have preceded him in the work laid out.

The handling of smooth rolls for the purpose of reducing
purified middlings to flour is one of the most important
operations connected with the conversion of wheat, and calls
for a display of intelligence such as is calculated to grasp the
system in detail, to know just how many times it ought to
be necessary to roll any given product so as to ensure of
its being finished without being injured or allowed to injure
anything else. It is the application of that inner practical
knowledge which can rarely be communicated from, or to,
the outside, and yet it is the result of study by book and
practice combined, built in, as it were, and applied with
more and more force the more the practice matures and
gains by its maturity, and it is in this application of book
learning that the author wishes the student to proceed to
this practical knowledge. When treating of the break
system it was explained that the rule was to allow so many
inches of roll surface (length) to each sack of flour made by
the mill's combined effort. It was stated at from about
35 in the cases referred to, and in this instance the same
calculation is applied to smooth roll surface, only differing
in the amount. As a rule it will be found to be sufficient
to remark that in the reducing of middlings sent to rolls
from purifier and other machines 50 inches per sack is none
too little. Some millers will require more, others profess
to be able to get what is wanted with 40 inches.
However that may be, the former number may be taken
as approximately correct for practical purposes herewith

connected, and it is upon this basis that it is proposed to treat at this stage ; and when the whole has been carefully digested it will not be difficult for the majority of readers to fit their ideas in with the general purpose engendered in the following pages.

It will be remembered that size has played a more or less conspicuous part in every development as yet introduced, and so it does here. The first rule almost will be that of running only material of the same size approximately to the same roll, as it is obvious that if a great, or only a moderate, diversity of sizes are allowed to travel to the same reduction roll, the work will be done in a very unsatisfactory manner. Either the roll must be set up to disintegrate the smaller particles or those of larger cubical measurement. If the former, then in order to reach the small cubes the larger are pulverised to such a degree as to render them a positive drawback to the quality of the flour, and if the latter plan be adopted the smaller slip through without being operated upon at all, and might with greater benefit have been left alone during that part of reduction, on account of the worming and tumbling action doing absolute mischief through over-travel, as pointed out in a previous chapter. There is a medium between the two extremes which generally is not very difficult to follow, and that is to allow no greater latitude than, say, 20 meshes of silk. To illustrate this we will take a case of feed coming from a purifier direct. The purifier, maybe, is clothed something like this—70, 64, 60, 56, 50, 44. There are six sheets, and in an ordinary case the first five will be found to be fit to run together to the same roll, both for quality and size ; the last sheet will be cut off to go lower down, and the tailings will be treated by themselves or with other tailings in a manner best adapted to their needs. By this means of classifying and grouping together what material is considered of equal size and quality, a shorter method is obtained and the power saved

is considerable, and this in addition to having a better flour
than would be the case were a more roundabout system
indulged in. A great many mills, no doubt, are handi-
capped on account of a lack of machinery in this respect,
while others have quite enough, provided the flow of the
different grades was more intelligently guided to the rolls
in better or more congenial company.

Stock of diverse quality also should not be sent to the
same pair of rolls if this can be avoided. The most glaring
faults in connection with this part of middlings reduction
are committed by millers of limited experience. As a rule,
the rolls are larger than the size of the mill or the quantity of the output warrants, and, this being the case,
there is not feed enough of one class to spread over
the whole of the surface ; consequently, two sorts of
feed are supplied—one for each end—but the result
is never satisfactory, either to the product or the rolls
themselves, and they are not long before the unequal strain
begins to show in the uneven wearing of the bearings and
pressure tackle, necessitating renewals of parts and
constant trueing-up of rolls to do anything like what
ought to be easy of accomplishment. This state of things
is partly brought about because of the less cost comparatively, but no doubt later considerations make it
appear false economy. The effect of rolling good and bad
stock together is that the latter is nearly always noted for
very small branny impurities, which mixing with relatively
pure material vitiate the higher results which would otherwise accrue from the separate treatment of each. It seems
on the face of it that purification is of little use if the
separated particles are to be mixed together again in any
form ; nevertheless, it is allowed to be done in some places
merely because after the first reduction there is not the
quantity of roll surface available, or it may be that given
a high-class first-product flour it does not concern the
miller to a great degree what becomes of secondary middlings .

so long as the maximum amount of first grade is produced.
But this is not what the young miller wants to know, and
the only reason for calling attention to it is that when
anyone comes to look at a mill's actual flow-sheet he is
apt to be surprised at finding it so different from what he
has been led to believe by a perusal of text-book literature.
The broad principles of milling are laid down in these
pages, and any and every variation therefrom on the part
of responsible millers themselves is made to suit local
trade requirements. I again assert that milling cannot be
learnt from books. They are only a landmark of general
observations, applicable in part or substance, and open to
countless additions or omissions to suit whatever class of
trade or local requisite may be deemed necessary by those
who are in authority.

Starting with the coarsest material from the purifiers we
shall have stock coming on to the first pair of rolls which has
passed through a silk mesh ranging between 40 and 24, or
thereabout. This is large semolina, and owing to its size
it is impossible that one roll will reduce it to flour in its
entirety; in fact, it is never attempted. There is a certain
pressure exerted upon it to reduce its size, and in doing so
to make what flour is possible under the circumstances; but
this is never a very great amount owing to the fact just
stated, and also because it would be inadvisable to do so
on account of the filmy substance attached to the several
particles making the first reduction roll yet another link in
the long chain of preparation for the production of the
highest class of flour. What is really aimed at on this roll
is the reducing of the semolina to a more convenient size,
whereby it can be sent to a second pair after the flour has
been extracted which was made in the operation, and this
second roll seizes the lesser sized particles, and makes
a short cut with them to the patent grade flour sack. Flour
made in reducing semolina to size is not of the best brand.
Experience teaches us that until we have the first hand

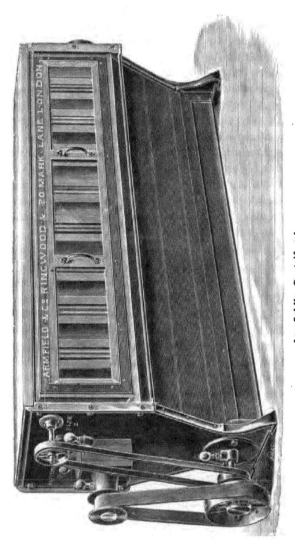

Armfield's Centrifugal.

middlings of a certain size we must not send the resulting flour to join the patents. I am speaking now of present methods, which are so arranged that semolina is, or ought to be, obtained as large as possible, reduced once, and the cut off employed to get the best middings from this arrangement sent to special rolls and dressers to be kept apart for special purposes. The greatest amount of skill required in working smooth or reducing rolls is in knowing when and where to apply or relieve pressure, so as to maintain an equal condition all round and under all influences. Care should be taken to see that squeezing is avoided. The differential speed of 1 to $1\frac{1}{4}$ or 2 to 3 provides for granulation. On pure stock a fair amount of pressure may be exerted to ensure a quick despatch, but if the residue is intended for re-purification this implies impurities, and squeezing or excessive crushing should be avoided until a thoroughly pure feed admits of severe treatment. Scratch rolls are employed in most mills to treat the overtails of the course purifiers. This is a part of the wheat berry which has small bits of bran attached and, being considered somewhat too large to receive correct treatment on smooth rolls, a separate pair are provided with fine corrugations and run at a higher differential speed. With 30 to 36 corrugations to an inch and a differential of 3 to 1 peripheral speed, the result is to shear away the bran snips and liberate the floury portion so that, after being dressed and re-purified, the bulk of it becomes fit to join the best runs of stock. Some flour has been made during the process, but this could not be avoided owing to the nature of the particles and their composition, and it is usually sent along with break flour or other lower quality according to trade custom and the requirements of the district. The severed bran snips are in reality germ particles which, after being rolled again on smooth rolls, are sifted over about No. 18 or 20 GG to get them entirely pure, and passed by themselves to a separate outlet and classed according to value, or mixed in the coarser runs of offal.

Before adjusting the rolls it is always best to closely examine the feed, so as to gain an idea how near to have them together, or, in other words, to avoid making the rolls do more than the condition and quality of the feed calls for. Compare also feed and product—that is, before and after rolling, because it gives one a keener insight into the needs of every kind of stock which may happen to be under consideration, and better results will be obtained ultimately, and that with less trouble on account of the educational value attached to it. Nothing beats cultivating habits of observation and carefulness—two qualities which go a long way towards making successful men in any department of life. As on the break system, so it is here in regard to the working of rolls as a whole, so as to make the results just what they ought to be—to harmonise the plan of reduction, doing a little more at one place than another, providing there is no flaking of the reduced stock. This will vary with different wheats or mixtures. Hard wheats make a goodly percentage of best middlings if properly handled on the break rolls, and which, if nothing were allowed for, the rollerman would find it extremely difficult to dispose of, and in that case a closing-up at the head of the reduction rolls would be almost imperative so as to get a greater pressure and reduce the middlings to flour before they accumulated with other less valuable stock further down the mill. Soft wheats, however, make fewer middlings, but they are not readily reduced to flour on that account. The roll pressure must be very slight, else the work will be badly done, the flour will not be so granular, the rolls will get pasty, and the work will be unevenly distributed over the rest of the reduction area. Put concisely, these two rules constitute the whole of the skill—viz., when and where to apply or relieve pressure so as to maintain an even distribution of material and so to work the rolls that nothing shall be done to injure the baking quality of the flour, and that all the flour there is in the feed shall be obtained in a pure state. The rolls should be examined constantly, as they are liable to vary in the work performed on account of heat—more or less—quantity and variation of feed as to quality and the other possible changes which only a rollerman who constantly watches them can become cognisant of. If a man be thoroughly in earnest he will on no account neglect this duty, because of the insight it gives him of everything likely or unlikely to take place in the working parts of the machines committed to his care. Little details show themselves more quickly when a man is alive to his duties. If the rolls are of the belt driven type, then the differential will perhaps give occasional trouble; through the slipping of a belt, the roll may not be doing its proper work and a little extra pressure applied may make it worse. If a man be not attentive this will puzzle him for a time until the matter is forcibly brought home to him by the slipping belt coming off entirely. If the belt slips, the pulley face will get hot and this is an easy way of understanding the matter; then the question is: what shall be done? The attention insisted upon will tell him whether the belt is too slack or the roll itself unevenly adjusted— that is one end closer up than the other. Hot bearings will also denote this and also over pressure and as has just been noted attention makes a man wise in regard to the vagaries of machinery. It ought to be stated here that whatever goes out of concord in this matter of reducing wheat particles will generally be found to emanate from the rolls themselves, or at least to show itself soonest in the actual work of the rolls. Uneven or excessive feed is the cause of by far the greater number of little troubles, and this also comes from neglect in some department of the mill. Bad dusting sends a lot of stuff wrong and interferes with the proper disintegration of everything it comes in contact with as regards the material under operation. Patchy silks which take up the dressing surface on reels or centrifugals are answerable for something on this head, as well as underspeeded machines. The prime consideration running through this chapter is how to make flour. Everything has been done that can be done to have the different particles as perfect as possible. Everything has been prepared for a speedy despatch of these particles, and the question is how to get the flour, get it quickly, get it pure, and get it all, and by this a rollerman is judged to a great extent. It is in the nice handling of each pair of rolls that the work is accomplished—the fine medium, which can be seen by an experienced eye— the medium between extremes. It is no use to get down on all and everything in the sure hope that in that case you are bound to have the flour somehow, because there is then a great chance of having all the flour and a good part of everything else in the same sack. On the other hand, too lenient a roll pressure means a crowding of good stock on the latter reductions and a consequent spoiling of what is converted, together with a heavy sack of offal, implying a short percentage of the pure article. As just stated there is a correct medium which it is always safe to follow. Do not attempt to finish at one operation anything which passes over No. 3 silk, because it cannot be done except in the matter of tailings. Pure or semi-pure material dressing *through* the above number will stand a good chance of being finished at once, provided the pressure is rightly adjusted, for while there are, perhaps, some impurities present they will on the whole—being so small—flatten out somewhat and so tail over the centrifugal dresser, while the granulated middlings or cut-offs fall apart in atoms minute enough to get through the silk mesh clothed up to No. 12, and this is exactly what is aimed at in the actual practice of reducing middlings to flour.

Perhaps a word or two regarding the " feel " of the different products from smooth rolls may be of use to those who are not well versed in the handling of them. Semolina, when it has passed through properly adjusted rolls, will feel moderately soft ; there should be no flakes, and if carefully tested by finger and thumb sharp granular middlings will be discovered. This bears out what has before been said regarding the inadvisability of attempting to reduce such stock to flour at one operation. At reasonable pressure, such as will bring the material to the above state, the large granules are reduced or softened down, the product will show a fair amount of flour let loose, and what number of the larger pieces had bits of bran or germ attached to them will, as a rule, show up in a more or less flattened form. This is accounted for by the fact that the portion at, and near, the germ end of the wheat contains more or less oily substance, and instead of this being broken up like pure middlings, the differential or shearing action of the smooth roll is only just sufficient to lay it out flat, so that its final elimination is rendered very easy of being accomplished. A roll working on medium-sized purified middlings may be set up to a very fine point. There is not much danger of spoiling the product, providing an ordinary amount of intelligence is brought to bear in the handling of the adjustment gear. The flour is here waiting to be pulverised to the requisite size, and the opportunity must not be let slip. When it issues forth it will be quite soft and feathery, slightly warm, and only needing to be sent to the centrifugal to sift out 80 per cent. of the quantity operated upon. A few floury flakes will be observable, perhaps ; but the least damage is done by having them here than in any other place. On the other hand, the rolls must on no account sweat or become pasty, as that implies heat and subsequent condensation and a tending in the direction of making flour which is slippery and will soon turn sour, especially if it happens to go a long distance, or be kept a while, or be packed too tightly in the sack.

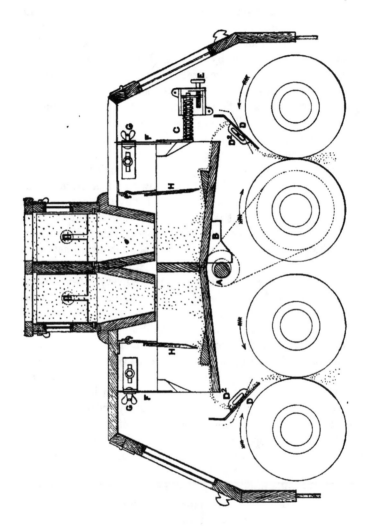

Robinson's Shaker Feed.

There is a safe medium in reducing pure middlings, and what that medium is I have just endeavoured to show. If the proper advantage be not taken at this point—if the rolls are not set up to the proper pitch—it is usually a source of loss in capacity, bad finish, and inferior high-grade flour. It should be borne in mind that this best and purest product demands more than partial reduction, on the ground that it contains the bulk of the nitrogenous or glutinous matter which, along with purity in colour, have both been brought here for the express purpose of being deposited as flour granules in a separate sack. The failure to reduce them properly means a failure in the patent flour, a consequent enriching of some ordinary grade which does not require it, a loss of percentage, and a monetary loss on the mill's wheat mixture. If the mill is running a straight grade only the argument applies with equal force, because a failure at this or any analogous point means a useless travel of feed, which goes farther down the system where it requires more power, and the general result is a loss of percentage in flour through being unable to get a proper finish, and the overcrowding of the later machines has a tendency towards making a slightly lower quality of material. Taken altogether this is intended to imply that wherever pure stock is found the opportunity must not be lost of reducing it as quickly as possible, without, of course, inflicting injury on any vital constituent. The writer has had considerable experience in this matter of best middlings reduction, and the rule among ordinary workmen is to under, rather than over, work the rolls, and that to an extent which in nine cases out of ten results in patents much below their proper value. Baked samples of flour from rolls operating but lightly on the feed of pure middlings show good colour but not much strength, while heavily pressed the results from the same feed and mixture of wheat give a loaf of good size and colour. This shows that in the first case the starch cells alone were operated upon by the light pressure, while with the pressure considerably increased the gluten cells, which are more compact and harder in structure, were made to contribute their share towards the ideal results. It is necessary that rollermen should understand this. The generality of workmen have been accustomed to look only on the finish, and if the finish shows nothing left on offal, and no offal in the flour, they think satisfactory work is being done, and that they need take no further trouble about it. This is by no means correct, because every roll is working upon some selected part of the wheat berry, and what that part is and how its composition is built up should be made a special study by those who would hope to be in the forefront.

There is a finer product we call dunst, which is also a lighter product than middlings, and looked at critically will appear to be that part of the berry which has encircled the starch and gluten cells—a sort of tissue covering. It is not so brittle or compact as the inner portion just mentioned, and has a tendency when under operation to flatten out more or less. Certainly it does this under the first rolling as middlings, and before the inner portion has been removed; in reality this part gives first notice of excessive pressure by flaking, thus acting as a sort of safeguard for the better material mainly under consideration at the time, and constituting itself a kind of cushion to protect the more valuable properties. This being so it does not dress out freely, and so engineers have made provision for its separate collection, and named it dunst. Pressed in the hand it has a soft feathery touch, and a high differential speed coupled with light pressure best suits its nature when obtained pure. It will feel slightly warm, and there will be some flaking; it will show a large amount of reduced material as flour, and 70 per cent. of the total should dress out for mixing with the ordinary brand or with the straight run grade.

An intermediate quality of stock for reduction will usually

be found in cut-offs, seconds, or other kind of semi-pure
stuff; it will consist of small sized pieces of branny compound mixed with starchy bits of lighter middlings.   It is
impossible to keep on re-purifying, as the difference in the
gravity of the two sorts is very slight. The usual way is to send
them together, as they are, to a special roll and allow for a moderate pressure being exerted upon them.   They are all more or
less reduced in size by the operation, but the tendency is also
favourable to flattening out the branny portion, it being
tougher, and on the other hand it is most likely that the
shorter-grained starchy bits will be disintegrated at the
point of contact of the rolls, and so have the best chance of
dressing out through the silk covering.   The flour obtained
is not by any means the best even from a straight run ;
wetted up it will be found to die off a deeper yellow than is
at all consistent with the making of a good loaf.   This
yellow residue is in reality the dye squeezed from the
brown material mixed with the feed, and also the minute
bits of fibre which in the best of cases it is difficult to avoid
making to a certain extent.   There is not much fear of
flaking this sort of stuff; the danger lies more in the
direction of multiplying the minute specks by over-pressure
to such an extent that the colour of the resulting flour gets
a deeper yellow or approaches a brown, according as the
rolls are regulated.   The judgment exercised should aim at
setting them just at that point which, while acting gently
on the starchy portion, will not shear too freely the other
and perhaps the larger part of the feed.   How this is done
is not within the scope of any writer to determine.   The
only thing to be counselled is constant and close oversight
on any particular mixture or single milling sample.   The
variation in condition, dress, system and quality of flour
cultivated makes it perforce a matter of individual
application.

Anything which will not pass through No. 40 GG. I
class as rough tailings, except from semolina purifiers, and

these call for a few words in relation to the finishing stroke which is applied to them previous to their deposition in the sack as shorts, coarse pollard, or sharps.   Speaking in a general sense, the heaviest pressure is applied here in order that nothing may be missed in a material noted for its tenacity in sticking to any adhering substance, and the nature of the stock itself, rendering some rough dealing necessary.   As elsewhere, the amount of flaking will be the guide to the limit of power applied.   Small pieces of germ, irregular bits of broken bran, everything which was a little too small to pass the first testing of the germ stock has by this time reached the tailings roll in order to be finished.

The rolls should not be allowed to get hot either here or elsewhere in the system, and this applies to the working surface as well as the bearings or journals, for, as before stated, hot rolls heat the stock, which then begins to sweat, and trouble is the result sooner or later.

I have already said that both vertical and horizontal types of roller mills can be recommended or otherwise, and I do not intend to further criticise their various advantages or defects; a reference to the chapter on the break process will be sufficient for satisfying millers and giving them the chance of proving the soundness of the conclusions therein arrived at.   The vertical type certainly gain a point in the matter of accessibility, and in some patterns this is more pronounced than in others.

It is very important that the rollerman should be in easy touch with every stage of the mill's reduction process, and the readiest mode of getting samples from the rolls undoubtedly attaches to those of vertical build.   A board made nearly the length of the rolls is a ready way of getting samples to guide the man in charge as to a correct alignment at both ends of the rolls.   Again, in the vertical pattern the scrapers are in full view and capable of nice and correct adjustment, which is more than can be said of the horizontal type.   This has always

Four-Roller Mill.—G. Luther, Ld., Brunswick and Liverpool.

G

been a serious drawback, as a great deal depends upon the rolls being carefully set, so as to keep clear of the flaking, which instantly follows should they be set or screwed up the least bit out of a straight line and cease to bear equally upon the whole length of the roll. Scrapers will also get troublesome when smooth rolls begin to show signs of wear and tear and get out of true cylindrical shape. Sometimes this is caused by uneven feeding or on account of the feed stock itself being of a different class at each end. It takes a bit of experience to locate the latter sometimes, the first idea usually being that the scraper wants re-adjusting and filing down straight or more pressure being applied to it. Both are wrong, however, and more screwing down of the scraper results in still greater wear and tear, more power is required, and hot rolls and hot bearings are the result. In the vertical type of roll the scraper should need no pressure at all ; the simple weight of itself ought to be sufficient to bring about a perfectly clean discharge of material without any other mode of friction, and so secure economy in the use of oil and a consequent cleanliness not obtainable except under good rollermanship. If rolls are not correctly adjusted at both ends there will be a straining at the bearings, and this will eventually lead to a distinct rattle in the gearing box, the rolls will work hot, and quantities of oil will be used.

I have sometimes tried to find a more convenient word than pressure, for, as a matter of fact, the rolls work at adjustable distances, which are somewhat less than the thickness of the material being fed to them, and the real pressure is caused by this material trying to get through a smaller aperture than its body will admit of ; hence it gets broken in the attempt. The only thing the rolls do is to withstand the pressure so created by refusing to let the particles through on any other condition, and if the stock happens to come out of the ordeal other than in correct form, if it should show signs of ill-usage, such as

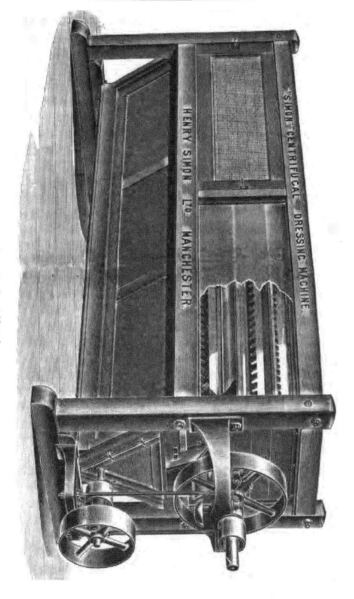

Simon's Centrifugal.

flaking, for instance, then the distance between the two rolls must be enlarged. If, on the other hand, it should be found that the particles operated upon have not been touched, or not disintegrated sufficiently, then the distance between the rolls is too great, and a consequent closing up is both advisable and necessary in order not to overcrowd any following pairs of rolls, as by so doing a good finish could not be obtained. Smooth rolls should preserve a dull frosted appearance, and this implies their highest efficiency in disintegrating whatever is sent to them in the matter of feed. This dull appearance is a sort of roughness which tends to hold the feed in order that the momentary grip at the point of contact shall be the more effectual in bringing about the desired result. On the other hand, a shiny or glassy cast on the roll is indicative of undue wear caused by short feed, uneven distribution on the surface, over-pressure, unequal character of feed, too keen a setting up of scrapers, or faults in the scrapers themselves.

A few words anent the reason why the difference in the number of revolutions of a pair of rolls is settled upon the figures as we know them may not be amiss at this point. Going back for a moment to the break rolls we will consider a pair as grooved 20 to the inch with a differential speed of three to one. The distance an individual piece of broken wheat would have to travel in contact more or less is very nearly half-an-inch. During this period the particle is held by the slow roll, while 10 corrugations (half-inch) on this roll's surface get past the point of contact, and free the particle from further treatment. During this period the fast roll, running at three times the speed, acts upon the particle with 30 corrugations—makes 30 cuts at it, in fact, and this difference, this speed relatively, has been found by much experience to suit the disintegration of the wheat berry best for the obtaining of broad flakes of bran and a large amount of middlings, accompanied with a small percentage of flour. To increase the ratio would cut up

the material too much ; to decrease it would flatten out the
stock instead of breaking it up. These are not actual
figures, but are simply put in so that the action of the
differential may be more easily understood.

In reducing middlings the case is somewhat different. A
less difference in speed suffices to give the desired result,
because a shearing action is not so much required ; it is
more of a crush with just enough difference in the speed of
the two to prevent its being a squeeze. The idea is that
the slow roll still holds the particle, while the fast one
operates upon it, but its passage is not retarded to the same
extent owing to the nearer approach to each other in the
speed of the two rolls, viz., three to two. The greater the
differential the greater is the probability that whatever
impure or branny matter is present will be sheared to
powder and so work its way to the flour sack, whereas if
the action, as just mentioned, partakes more of a crushing
nature the middlings are reduced and the branny pieces
being more fibrous and, consequently, more tough, are
more likely to be flattened out, and can thus be separated
by tailing over the dressing machine. This it is which has
settled the relative speed at three to two as being more
suited to smooth rolls than a larger or smaller difference.

The method of adjusting the rolls should be simple and
handy without having to use a spanner or other tool, both
ends of the roll should be capable of being adjusted together
as well as separately, and the lever which releases the
pressure should also stop the feed.

Tramming or trueing up should be practised at intervals
of a couple of months, and if the feed device be of the
shaker pattern one eccentric should not serve both sides in
a double set. It not infrequently happens that the feed just
fails to reach the extreme ends of the rolls, and the practice
of reducing or lowering about one-eighth of an inch at each
end has much to recommend it. Far better is it to do this
than to find after a few months of work that the ends are

shaping themselves into a minute ridge, and tending to keep the body of the roll from getting near enough to the material to give it a proper reduction, and a little—very little—slipping through of unreduced feed is far and away the lesser of the two evils.

In estimating the number of inches per sack of flour made per hour and apportioning them to the different qualities, the following tables will, perhaps, be found useful to those who encounter a difficulty in their attempts at making out-flow sheets :—

Two-Sack Plant.

Reductions	1	2	3	4	5
Inches of roll surface	20	15	20	18	15

Five-Sack Plant.

Reductions	1	2	3	4	5	6	7
Inches of roll surface	30	30	40	24	40	24	24

Ten-Sack Plant.

Reductions	1	2	3	4	5	6	7	8	9	10
Inches of roll surface	60	60	60	32	32	32	32	60	32	32

Fifteen-Sack Plant.

Reductions	1	2	3	4	5	6	7	8	9	10	11	12
Inches of roll surface	80	80	80	40	40	40	40	80	40	40	40	40

Twenty-Sack Plant.

Reductions	1	2	3	4	5	6	7	8	9	10	11	12
Inches of roll surface	80	80	100	80	80	80	80	100	80	80	64	64

If larger plants than these are erected the writer recommends that they be in duplicate. Two 12-sack combinations are better than one large 25-sack plant, and, with a little allowance all round, the two smaller ones would do the work much better; they are not so unwieldy, and should anything occur to cause a stoppage the advantages must be obvious. Besides this, a better division of stock can be made by having more machines of less capacity. The first cost, no doubt, is greater, but loss of time is avoided and the work at all hours is better under control.

It will be noticed that the greater number of inches are allowed first, on the large middlings and semolina, and second, on rolls which are supposed to be working towards a finish on rough tailings. Both require as thin a feed as can be managed, one owing to size and the other to toughness. There is a general consensus of opinion that rolls, whether for break stock or reauced middlings, should be about 9 ins. in diameter, except for very short or very long ones. Up to 15 ins. they may average 8 ins., and over 36 ins. 10 ins. is not considered too much, adding as it does a greater resisting force, required where heavy work has to be done and the journals are so far apart. Small diameters require a wider differential speed; the curve is so acute that to get a satisfactory amount of work done on each pair a greater influence must be brought to bear, and the only way to do this is to increase the difference between the fast and slow running rolls. On the other hand those rolls which exceed 10 ins. in diameter may, with benefit, be brought nearer to each other in the number of revolutions each has to perform. The reason for this is obvious, on account of the greater length of duration of contact arising from the exposure of a greater amount of roll surface when at their nearest point to each other in actual work. To put it plain a particle in the latter case is seized upon sooner and held longer than with smaller rolls, and if the same ratios were maintained the shearing action would be excessive and spoil the best part of the work. The ratio of differential speed for smooth rolls has been found to be somewhat as under, when taking that recognised as best for a basis : If a 10-in. roll revolves twice and its fellow three times, a 12-in. pair to do the same work would require a difference of $2 \cdot 2\frac{4}{5}$, a 14-in. pair $2 \cdot 2\frac{10}{22}$ and a roll 16-in. diameter $2 \cdot 2\frac{1}{3}$, thus gradually reducing the difference in speed the larger the surface curve exposed to momentary contact.

As part of the general plan it will be here necessary to

give a few concise rules for the guidance of rollermen, which are calculated to form a handy page of reference at all times:

1. Stock to feed hoppers to be even in quantity, quality and size.

2. Graduated pressure, according to size and quality of middlings.

3. Heaviest pressure upon tailings to ensure a good finish, and upon pure fine stock to speedily reduce to flour.

4. Careful attention to thin regular feed, and avoid flaking good material.

5. Exhaust enough to be applied to carry away the generated warmth, but not so much as will-draw off any flour.

6. Attention to scrapers to keep them level and to make them touch the rolls gently over their full length.

7. To allow of no escape of unreduced stock through faulty casing or other avoidable causes.

8. Avoid rolling stock too many times. When finished send to offal direct.

9. Equalise pressure, so that what is tailed over dressing machines is really finished, and reduce sufficiently at the head of the system to avoid overcrowding on the latter rolls.

10. Exercise a correct judgment by not sending a low-class flour to a high-grade sack, or anything which is really of a better quality to a secondary brand. Knowing the destination of the flour from every reduction, strive to get just that quality out of the roller feed which the designer of the flow-sheet intended should go there, no more, no less, and so keep the standard of each quality regular.

11. Give attention to the sufficient lubrication of bearings, run off the spent oil, keep the moving parts clear of dust accumulation, and so prevent

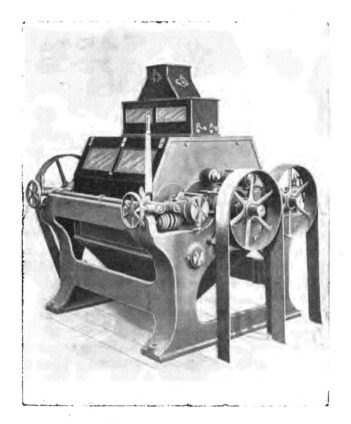

Robinson's Double Horizontal Roll.

undue wear of parts of the machine's mechanism, all of which tend to prolong the life of the machines, and save trouble and expense of renewals.

12. Driving belts and pulley faces to be wide enough to take any little excess of feed without interfering with or losing the differential speed.

13. Keep the rolls parallel to each other and in a correct horizontal plane or position, as middlings will pass through an opening $\frac{1}{250}$ part of an inch wide unreduced.

14. Take up any slight wear and tear in the brasses; this relieves the pressure on the spring connected with the adjustment tackle on most rolls, and gives it an elasticity it would not otherwise possess, and also allows more freedom of action should objectionable screws or elevator bolts find their way to the feed.

15. All feed-gate springs or other automatic devices, all feed rolls and feed hoppers to be kept free from moth creations or other obstructions likely to interfere with the regularity required at this point.

16. All fittings, screws, nuts and bolts, set pins, caps, keys, levers, handwheels, fulcrums, and rods to be examined constantly to avoid more serious happenings.

17. If there are facilities, semolina or any of the larger sized stock of value is capable of being reduced with less differential than smaller pure feed, such as twice purified middlings or dunst, which latter will bear slightly more shearing. Tailings also will stand it and be better for it, but as the market value of offal generally varies in all parts, the matter is best left to individual judgment.

18. A regular examination of stock as it *leaves* the rolls for the dressers is advisable to obtain judgment in the clothing of silk machines.

19. Scratch rolls will require delicate handling, the high differential (8 to 1) and the fine groove (30 to 36) being capable of a lot of damage if not properly attended to.

20. Do not shear or squeeze large brittle material but give more of a crushing power, and do not crush rough fibrous feed, but shear, as any other treatment squeezes out inherent or applied moisture, pastes the rolls and cases, tends to lower the quality of whatever is obtained, and blocks the whole mill by choking up the spouting.

Some millers argue that the best flour is obtained from the first reduction, others make the second their pet machine, and, again, the third has perhaps the most friends of any. If by first reduction is meant the first and largest semolina, then, I think, they are altogether wrong, and the reasons have been given in this chapter why this is so. If, on the other hand, the first reduction is so called because it is working on the smaller pure middlings—middlings which have been stripped of their outer coating of starchy matter and the filmy tissues attached thereto—and is intended for the highest class of flour, then they are right. Owing to size the first and second reductions are not complete and the flour is not sufficient, either in quantity or quality, to justify its being so labelled, simply because the roll is first on a flow-sheet. A much more perfect result can be obtained from the third reduction as the right sort of stock is there—it is small and pure, it contains the most gluten along with every other good quality there is in the mixture, and its uniformity lends itself more readily to treatment. As a general guide the following may be taken as a sort of rule regarding the qualities of the various flours from the machines comprising a 10-sack plant erected on present flow-sheet lines :—

1.—3rd reduction flour.

2.—2nd   ,,   ,,

3.—1st   ,,   ,,

4.—4th   ,,   ,, from 4-break reducing middlings
and purifier cut-offs or tailings.

5.—3rd break roller flour.

6.—5th reduction flour.

7.—4th break roller flour.

8.—6th reduction flour.

9.—7th   ,,   ,,

10.—2nd break roller flour.

11.—Remaining reduction flour.

12.—1st break roll flour.

13.—Bran flour.

The whitest flour (and perhaps the weakest also) will be
found on the second reduction dresser, which is obtained
by cracking the larger sized throughs of purifiers free from
starch, and so getting them ready for parting with their
strength on the next roll. To give another illustration
regarding the precise manner this comes about we will
assume—which is more or less a fact—that the dull frosted
or semi-porous surface of the rolls has an affinity for the
particles of semolina being acted upon. As they drop on
to, or are carried close to the point of contact one roll is
busy holding the particles back very slightly, while the
other and faster roll is wishful to get the operation over
quicker, and therefore each side of the semolina receives a
slight drag, one up, the other down, and this results in the
inner position falling untouched to be acted upon for its
own sake at the next operation. Assuming this it will be
easy to see that if the rolls are set at the right pitch the
law governing the operation is fulfilled and the after
results guaranteed, but should the aperture be too large
then the particles do not get enough of the opposite

drag to disintegrate them sufficiently, and the highest point is not reached in patent middlings.   If the rolls are set too close to each other the squeezing becomes accentuated, the purer inner portion is injured by being flattened, and what was intended to be broken away is found still adhering in a more or less flattened form, leading to the same result as the other, viz.: a depreciation in the patent flour. The work will also ·be longer under operation, and thus take up additional power and crowd the remaining rolls. It is said that genius simplified is nothing more than having an infinite capacity for taking pains, and I am of opinion that there is a good opening for displaying this characteristic in connection with the reducing of properly made and properly prepared semolina, middlings, dunst and other wheat products to their most valuable equivalent in flour, so that nothing is either lost, sent astray, or in any way injured during the process.

Another way of expressing the value of reduction stock for flour, is by saying that for first patents the purifier middlings once reduced and two runs of selected dunst will give the most all round satisfaction. For patents of a secondary character the remaining three or four dunst cutoffs and re-purified middlings form the staple quantity, and for bakers' quality (generally) rejected middlings, four lower reductions and the third break flour ; the residue of the mill being classed as seconds.

Yet a third way of doing this is exemplified as follows:— First patents from middlings passing through GG from 70 to 40 on the purifiers ; second patents dunst from first group of dressing machines and (say) two more purifiers from 60 to 32, both these to be rolled and dressed.

A third grouping would include purified stock from the fourth break, re-purified overtails, and the flour from the second and third breaks. A fourth quality would include what remained of the system except a cut off from the

last centrifugal through about No. 9 silk, and this joined
by the first break flour would constitute low grade.

The flour *par excellence* is obtained from the middlings
of the second and third break machines, through 40 GG
and over No. 6 silk, and purified through 60, 50, 46 and
40. This should be rolled with a fair pressure, and,
dressed alone as a product, it will be found to be the best
in the whole mill. As is previously noted the reason for
its being so is that according to our latest knowledge and
system the machines mentioned both here and before are
supposed to operate upon that part of the wheat berry containing the bulk of the gluten combined with high resulting
colour in the flour and bread according to the physical and
chemical construction of wheat as understood for flour-
making purposes.

For the reader's sake it will, perhaps, not be wise to
enter further into the question at this stage, but to leave
the consideration of wheat characteristics until a later
period as more calculated to form a sequence in the order
the book was entered upon. A few words anent the precise
meaning of differential speed may help to engraft the
subject more firmly in millers' minds than the mere
mention of such a thing being necessary is likely to do.

Differential velocity refers only to the circumference.
It means an absolute difference in the number of feet
travelled by the circumference of two rolls in the same
length of time. It does not in any way refer to the varying
velocity of the revolutions of any two rolls working together,
because any two such of different sizes may run at different
speeds and yet not have any differential velocity. As an
instance of this we may take a roll 30 inches in circumference and making 300 revolutions per minute working
in conjunction with another roll 20 inches in circumference
and 450 revolutions in the same period of time. In this
case there would be no differential velocity whatever, as 30
inches and 300 revolutions per minute equals 9,000 inches

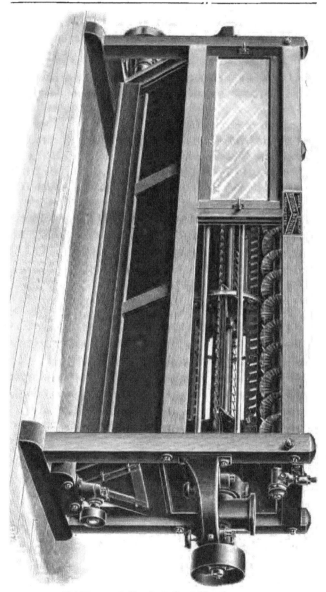

Briddon and Fowler's Centrifugal.

travelled ; so does 20 inches and 450 revolutions per minute equal 9,000 inches travelled.

The result of having no differential has been already explained as shewing a crushed and unshapely mass of material, and upon which greater pressure would have to be exerted ; consequently more horse power would be consumed, the wear and tear of machines would be considerably augmented, break flour would abundantly predominate, middlings would be almost entirely absent, and the whole system would be thrown out of agreement on account of being underfed with inferior stock. The flour would be soft, silky, and lifeless, and a close finish of offal would be totally impossible within a reasonable limit of time, expense, and machinery. The writer is certain that after this repetition as to the necessity of proper equipment and treatment the subject can be well left to mature with increased experience.

As another example in explaining the meaning of differential velocity it may be said that speed or rate of motion is simply measured by the amount of change of place or displacement in any given time, and the direction in which this change takes place is properly specified. In connection with the subject under discussion, viz. : circular rolls, the velocity is the distance travelled within the boundary line of the circle. In the case of rolls running in pairs and close together, the differential speed is the circular distance travelled by the two rolls in a given period of time, and has nothing whatever in connection with the numbers of revolutions, but where we have the same sized rolls the differential speed may be worked out as follows. In the case of two rolls 10 inches diameter, one making 150 revolutions per minute and the other 350 revolutions per minute, the differential speed would work out as follows :—

$$V = \frac{150 \times 10 \times 3 \cdot 1416}{10 \times 22} \qquad V = \frac{350 \times 10 \, 3 \cdot 1416}{10 \times 22}$$

$$V = 150 \times \frac{10 \times 22}{7} \qquad V = 350 \times \frac{}{7}$$

$$V = 4{,}710 \qquad \qquad V = 10{,}990$$

Roll travels 4,710 inches   Roll travels 10,990 inches
per minute.   per minute.

As 4,710 : 10,990 so is the differential velocity.
This is as 8 : 7 (or as 150 is to 350).

But supposing that the rotating of one roll is to the other roll inversely as the circumference of one to the other there would be no differential at all, so that as previously noted two rolls may run at different speeds and have no differential if the circumference of one is different from the other. But if, as the above example, we know the different sizes it is a simple matter to find out the distance travelled by each.

In writing of a certain number of reductions it does not follow that the whole of the middlings must pass in consecutive order through each and every pair, and are not to be considered finished until they have done so, nor is it intended to be conveyed that there are a dozen different sorts passing continually from the purifiers to the rolls. The best kinds naturally are treated alone, others are mixed with a similar quality from somewhere else, while the lowest quality after separation may be, and is, sent direct to a finishing roll. The knowledge gained by studying the various runs of material from this standpoint is indispensable through a miller's life, and tends to make him an expert in classifying every thing that comes through his hands in the shape of roller mill feed.

Smooth rolls, quite as much as break rolls, require to be kept true, and if the system is a short one there is all the more need of watchfulness in this respect. If a pair of smooth rolls, for instance, become the least bit out of truth, or only one of them so, an opening is made when the rolls are brought together at that point of their circumference.

If one or both happen to be " full " at any point, the tendency is towards pressing the remainder of their surfaces apart, and this will allow material to pass through them untouched. On the other hand, if there are patches which are "low," spaces would be there even when the rolls were brought together, and again reduction is not accomplished in a proper manner, so that it really does not matter what form the defect may take, the result is the same, viz., an escaping of material farther down the mill. If it is a

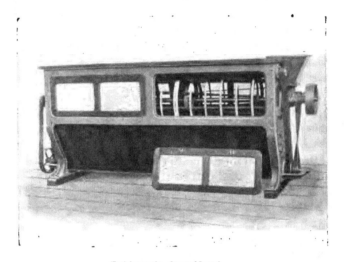

Robinson's Centrifugal.

roll at the head of the reduction series which fails in this respect it tends to overburden those rolls which come after, and even if the latter are in good condition in every way they get more than their capacity admits of treating aright, and so are liable to do it imperfectly; the mischief, in fact, grows as it gets shelved again and again until the total capacity of the plant is considerably reduced. A good way of finding out these imperfections in smooth rolls is to bang them up as close as possible, place a light inside the framework immediately level with the line of contact, and if the light can then be seen either the rolls are not close enough or they cannot be got closer. If the latter, then turn them round slowly, and by and by the mischief will be self-evident if the line of light be very carefully watched. If the light can be seen at any time during a revolution there is some fault at the point of discernment. This is a very simple test, and also about the best available away from a lathe, as the smallest defects will thereby be revealed. The writer is glad to be able to say that, through a long experience, rolls fresh from the makers are seldom or never found to be defective in this respect. Years ago there was a difficulty in getting the metal uniform in texture, there was a mixture of hardness and toughness, but that has long ago been overcome, and now it is seldom thought necessary to go through any elaborate inspection when setting fresh or new rolls to work. The danger is in running them too long afterwards, and money is lost indirectly in continuing to work smooth rolls at a pressure when that pressure does little or no good through uneven surfaces being constantly wedged one against the other and made to revolve at a high rate of speed. As was just mentioned in the chapter on breaks, a difference of opinion exists regarding the speed of feed rollers, some millers maintaining that they run far too slow, and, while not going so far as to say they are wrong, I would point to another aspect. Middlings, we by this time know, are, on the whole, very friable, and everything possible is done to avoid violence when handling them, so as to prevent disintegration taking place prematurely. Feed rolls are grooved also to allow of the stock being taken past the inclined gate with as little friction as possible. Some resistance, however, is always offered, and no one can say that a slight breaking up of the particles does not occur, and if that is the case their acceleration of discharge means more frictional contact, and consequently

more irregular material, and though only perhaps to a very
small extent does this take place, it proves that to increase

Turner's 4-Roller Mill.

what is already a negative fact is wrong both in theory and
practice.   The shaker feed, however, is free from this

suspicion, and perhaps that may have something to do with its continued usage. To many millers the subject presents itself in another light. They argue that it is not so much a question of quick delivery from the feed hopper, but it is more important to know what speed it enters the reducing area or point of differential contact and to place the stock to be reduced upon the surface of the fast or highest speeded roll is a more correct way of doing it, and that without friction at all provided a shaking device be employed to distribute the feed in an even and regular stream.

It is quite clear that as middlings fall upon the larger and quicker revolving surface they will be thinned out considerably and thus, to an extent, fulfil this law of gradual reduction milling. That, further explained, means a very thin sheet of feed travelling at the same rate as the fast roll, and all it needs for being reduced aright is the check it receives when coming in contact with the slow roll. It must always be understood that rolls are not supposed to be absolutely closed up against each other, and what slight opening there is between the two regulates the size the material should be after being operated upon by them.

Hard, true rolls, noiseless and easy running coupling gear, long bearings and well-fitting accessories for the same, rigidity of roll casing and floor foundation, eccentrics in perfect accord, fine screw at each end easy of manipulation, scrapers of soft mild steel, which, as a consequence, will not wear into the roll's surface, every working part easy of access to regulate, differential speed, whether by wheels or straps, and all the other items which strike a thoroughly observant rollerman to be kept up to the required pitch of excellence—these are the principal points. Everything, so to speak, depends on everything else, and nothing can be the least out of line without affecting something else, and in so far as these little points are not attended to, by so much will the mill's efficiency be interfered with.   Little slips have far-reaching influences, a fact easily proved by watching the result of any such right down the mill's flow-sheet.

Reviewing this chapter, it may be stated that the following questions bear more or less on the subject matter, and should be fixed in the minds of students for future usefulness :—

1.—What is the main object relating to the reduction of middlings ?

2. How are middlings divided and sub-divided?   Give reasons for doing it.

3.—What are semolina, middlings and dunst ?   About what number of silk will they severally pass through or over to merit their names, and where do they mostly abound ?

4. What would be the result of sending large and small stock to the same pair of rolls ?

5. What causes flaking ?   How can it be avoided, and what follows, if persisted in, or allowed to continue, on stock which is lively and granular to the touch, where it is most difficult to avoid, and at what points is the mischief done of the least and most importance ?

6.—Why is it considered best that stock of even quality should be kept from being mixed either before or after rolling with stock of a different character?   Some experts argue that fine and coarse material are best dressed together after being rolled separately ; give reasons for, or against, this practice.

7.—What are cut-offs ?   What purpose do they serve, and what is generally done with them, both on purifier and centrifugal or other dressing medium.

8.—When is it advisable to exert more pressure at one place than another?   Give two instances and specify reasons for so doing.

9.—What is the result when smooth rolls are either over or underfed on large, medium, and small middlings, dunst, and tailings? How would it be best to treat feed one-eighth of an inch in thickness?

10.—Where are the best middlings to be found, and what constitutes their special qualifications for being thus named? How should they be dealt with on smooth rolls?

11.—Why is it necessary that an even feed, the full width of the roll, should be constantly maintained?

12.—What are tailings? Why are they so called, and what part of the wheat berry is responsible for their presence? What increases their quantity, and what state should smooth and break rolls be in to ensure the minimum quantity being made?

13.—What would be the results of rolling material too long and not long enough? Describe the correct medium.

14.—Smooth rolls differ from break rolls in relative, peripheral, and differential speed: why should that be considered essential? What is about the correct figure for the former, and what would ensue if there was no difference in the revolutions of a pair of equally sized rolls working upon material in a mild, hard, or soft condition?

15.—What is the rule as to scrapers? How should they work or act, and what should result from a right performance of the duty allotted to them?

16.—Describe a simple feedgate device and say how the movable parts and the parts at a tension are affected by an inrush of feed, or a shortness of the same.

17.—Does the speed at which the feed enters the rolls proper have anything to do with the rate of reduction as to quantity?

18.—How are scrapers best attached to the framework of rolls? Describe the parts of those employed on vertical and horizontal rolls, and state in what respect they differ.

19.—What is " tramming " or trueing up ?   How is it usual to proceed to ascertain if rolls are out of cylindrical form or parallel adjustment ?   Give three or four ways of discovery.

20.—How should reduced semolina, middlings, dunst, and tailings feel when handled, and what is the guide to say the latter are finished ?

21.—State points for and against vertical and horizontal types of rolls for reducing purified stock to flour, and give reasons for and against belts and toothed gearing.

22.---Give your opinion as to the most suitable incline of spouts the various ingredients forming smooth roller feed will best travel from the horizontal, always avoiding the vertical position.

23.—What is the primary reason for engineers allowing (say) a minimum of 40 inches smooth roll surface per sack of flour manufactured in this department of the system, and how would it be apportioned in a 5, 8, 12, and 20-sack an hour plant ?

24.—Name all the parts connected with a set of rolls from the hopper to the floor, both stationary and movable. Describe the action of the latter and the necessity or function of the former ; take all sizes and speeds and make a memorandum of them for future reference.

CHAPTER IX.

Flour Dressing and Flour Dressing Machinery.

One of the most important operations in connection with the manufacture of flour is the dressing of the rolled product, and this is effected in various ways. Formerly the miller was perfectly contented with a sharply-inclined barrel, into which the whole meal was fed in the same state as produced by the stones —viz., without any previous separation, without any desire towards extracting what we know as middlings, but with every particle of the pulverised wheat intact. This primitive machine was covered with a wire cloth, representing, perhaps, a fineness equal to No. 6 or No. 7 silk, and the throughs would be the then flour of commerce, representing fully 75 per cent. of the whole wheat berry. After a time woollen cloth was employed, and finally silk came to be recognised as the best medium for the purpose.

Upon the introduction of the roller process things underwent a change in the shape of centrifugals, and it says very much in their favour that for a long period they held the field in the flour dressing department. Looking back to what may be called ancient milling history, we find one, Gustav Lucas, a milling engineer, of Dresden, built a machine with moving cylinder and revolving internal beaters in 1861. It was a long time, however, before this machine met with general approval among the milling fraternity, and although a full description of it appeared in "Dingler's Polytechnical Journal" in 1862, only a few

machines were sold during the succeeding five years for
flour dressing purposes. It found more favour with the
manufacturers of chocolate, sugar, and various powdered
drugs, and it was not until 1870 that Messrs. Nagel and
Kemp took it up, with the result that it quickly became
popular, and the thousands at work to-day testify to the
machine's efficiency.

It is quite a different process we are now about to
attempt to explain, as, contrary to the earlier procedure,
the main idea is to extract every particle of flour supposed
to be let loose by the action of the differential and
peripheral speed of the smooth rolls, and in every way to
facilitate the deposition of these particles in the flour sack,
and that in a pure state. In order to correctly understand
the process, it will be advisable to follow the run of the
stock through one of these machines and see in what
manner it does the work allotted to it. Considering that
the long line of preparation has been successfully passed,
there is no need for that extreme gentleness hitherto
preached in regard to the handling of the product.

The centrifugal is provided with a set of internal beater
blades which revolve at a speed of some 200 revolutions
per minute inside the silk covering. These beaters are
3 or 4 inches wide, and extend the full length of the
machine. They are usually set to revolve about 2 inches
from the silk's inside surface, but by means of slots they
can be either set up or set back according as the stock
being operated upon demands it. They are also set more
or less at an angle, that is, in a slanting oblique direction
looking towards the tail end, so as to facilitate the passage
more or less of the material being fed to them. They
differ slightly also in construction, seeing that some have
perforations in them, others are cut into strips about half
their depth and bent slightly, while a third variety present
a smooth surface the whole length. Eight to 10 are
considered the correct number; more than this number is

said to tend towards keeping the feed about too long;
less than this number throws larger quantities against the
silk's surface with too great a force, and hence frequent
breakages are brought about and the life of the cover is
considerably shortened in consequence. They are keyed
upon an internal shaft running the whole length of the
machine, and which as a rule constitutes the main driving
apparatus. The primary object in their construction is to
enlarge their capacity over reels pure and simple, which
they do in about a ratio of two to one, and also to disintegrate any part of the feed having a tendency towards
flaking under the operation of being reduced by smooth
rolls. The feed is introduced by means of a short screw,
and on its entry is supposed to be caught by the revolving
beaters and hurled at the silk cover with some force, and
what is small enough to pass through the meshes falls
away to the collecting worms in the shape of flour. The
beaters also are supposed to keep the stock away from the
heart of the machine and continually against the silk, at
the same time gradually working it towards the tail end
by means of the spiral twist given to them. Setting up
the beaters nearer the silk increases the frictional action ;
setting them back in the slot does the opposite ; the nearer
these blades approach the horizontal, the longer the stock
is kept in the machine, while an acute departure from the
same ensures a more quick despatch of whatever happens
to be operated upon. Centrifugals are either round, six,
eight, or ten-sided ; each has something to recommend it
to users, and all are alike built to get rid of as much stock
as possible in the shortest time consistent with good work.
A brush is attached on the best patterns, which is chiefly
found at work somewhere near the top or just over the
travelling side, where the supply of material is found to be
lightest, thus giving it a better chance to clear the meshes
at each revolution than it would have if it were placed in
any other position.

The revolutions of the machine itself will be from 25 to 35 per minute, and the numbers of silk through which the particles of flour are forced are mostly confined to 10, 11 and 12, ranging from 108 to 124 meshes per inch in length. Nearly all centrifugals are fitted up with what is termed a cut-off sheet, which is intended to cut off or divert some portion of the product after the bulk has been freed from whatever quantity of flour particles it possessed. If these cut-offs are intended to go direct to a succeeding roll they are taken through a comparatively fine number of silk, say 5 to 8 in the table given. If, on the other hand, it is considered best to cut off a product for purification or re-purification, then a coarser number may be employed, so as to make sure of having all that is thought to be of the requisite value to undergo the process, and when this is done the residue is called tailings, and is usually sent to a finishing roll away from stock of higher value. It may here be stated that in the arrangement of flow-sheets this matter of cutting off and the disposal of the product so treated calls for the highest knowledge a miller can possess, because it depends, to a very large extent, upon the nature of them, where they are sent, what brand of flour the residue joins, and it also relates to the finishing up of everything well within the mill's compass of machinery, without crowding, and without short feeding any apparatus in connection therewith, and a man not possessing a thorough knowledge of the construction of a grain of wheat, physical and chemical, should never attempt to interfere with existing arrangements until he knows exactly how any alteration is likely to interfere with the mill's product as regards any given brand of output. As a general rule it may be said that dressing flour is mainly a matter of separation by size. It may also be stated that cut-offs from large valuable stock is for purification, and from smaller, purer feed a dunst sheet is the medium between two sets of rolls. Large semolina will

Turner's Centrifugal.

stand three reductions. That is, one for sizing down, a
second for getting 80 per cent. as flour (after second
purification), and a third to finish it off. Ordinary
middlings should be about finished at twice rolling,
except what may be left over in the shape of light tailings.
Dunst should show very little after' being thoroughly
rolled and dressed, although a small quantity of light
filmy tissue may remain for the last reduction. All
machines having overtails of a branny nature should be
provided with a sieve to relieve the following machinery,
and all rolls acting solely upon tailings should also be so
arranged that any finished stock is prevented from entering
the flour-dressing machines. Seeing that the dressing of
flour is also a hunt after colour, it should be so managed
that stock of diverse shades are never allowed to be mixed.
It does not matter to the machinery under notice what the
wheat mixture is or the quality of flour demanded in so far as
regards other attributes, but colour must be a prime object,
and to get it every device must be employed which suggests
a possibility of increasing the whiteness and purity of
the flour.

The size of centrifugals ranges between 18 ins. and
36 ins. in diameter, and the length of them from
40 ins. to 120 ins. Two worms are best, fitted with
adjustable flaps so as to examine the dressing of every foot
surface and direct as seems best to suit. The rapidly
revolving beaters cause a current of air, and it has puzzled
many young millers to ascertain how this is obtained and
where it goes to. As a matter of fact very little goes
through the silk, and there is no downward or upward
draught from the tail delivery, so that the only conclusion
arrived at is that the machine draws in at the centre what
it delivers at the circumference without troubling outside
atmospheric density, which is equal to saying that it uses
the same air over again continuously.

It will be in the recollection of millers that a few years

ago there were several other devices on the British market for dressing flour by the aid of sieves. I have already mentioned the one type which has survived, and we must now take note of others which have lately been put before the millers of this country. The firm of Messrs. Amme, Giesecke and Konegen is by this time familiar to all who take an interest in trade technics and trade usages, and it only remains to be said that they are now erecting complete mills on the newest plansifter system, and the writer is informed that very high-class results are being achieved with less total machines, taking up less floor space, and at a considerable saving in the total amount of horse-power. I am also told that the latest type of machine is different in every detail from those of a former period, and that no matter what the wheat mixture is, or the class of trade cultivated, plansifters can be made to suit the situation. It is apparently the rule of this firm to build according to individual wants, and this is no doubt a great feature in any business.

The Plansifter under notice is built on the simple and effective principles of out-balancing. The sieves are arranged in two separate dressing chests, each independent of the driving gear. The flywheel is situated in the centre.

The frame into which are fixed the dressing boxes is of solid wrought iron, and the flexible projecting ends hold the chests securely and firmly. The machine is driven from below, thus avoiding guide pulleys and keeping the sleeves accessible by reason of the absence of any belts and pillars. The inlets and the overflows of the sieves all occupy the entire width of the same. The discharge spout for the flour of each sieve extends over the whole length, so that the flour, immediately it gets through the silk, is practically delivered into the spout. The silks and wires are kept clean by brushes.

By using these sifters, which practically represent scalper, grader, and flour dresser in one machine, less floor space is said to be required than would be necessary in the case of other machines, and at the same time all intermediate spouting between scalpers, graders and flour dressers is practically avoided.

The dressing of flour through centrifugals is an easy matter provided the material is rightly reduced, good in condition, and suitable covers are employed. It is when the reduced middlings are the opposite that difficulty is experienced in getting the best results. Wheat, which is out of condition, will yield middlings that after passing through the rolls will cling together, will be to an extent flattened, and the capacity of the centrifugal will be inter-. fered with to a degree governed by the departure from normal conditions of work. A change of mixture containing more hard or soft wheat than usual will show itself here as well. If the change is from hard to soft a coarser system of silk covering will be required to avoid sending some of the flour forward through the cut off, and this repeated several times will find the last machines over-crowded with the wrong quality and prevent a close finish up, and the least harm will be done by lowering the quantity of feed in the shape of wheat going to break rolls. If the change is the other way and atmospheric conditions are favourable, the mill will carry an increased feed, and should the alteration be not too extreme it may be sufficient, and so avoid the necessity of changing silks. To make this matter more plain for young readers we will imagine a mill working a single hard wheat (Indian) in its natural state, viz.; dry, brittle, and short grained. To get this into flour requires less machinery than a selection of English wheat. The former falls to pieces readily, is easily purified and rolled, and in the dressing department very quickly finds its way through the mesh and into the flour sack. Contrast this with the working of the latter, where every operation is a difficulty because of the clinging nature of its particles, their liability to clog, and the danger

Plansifter.—Amme, Giesecke and Konegen.

H

of spoiling the whole proceeding by over-pressure or other
violence noted in a previous chapter. It is easy to get rid
of a handful of bullets in comparison with a handful of
feathers, and although the simile may be crude, the gist of
the matter can be discerned therein.

The following is a table of silk numbers and will be
found useful in every-day practice. It should be committed
to memory in its entirety for reasons which ought not to
need any explanation. Contact with silk is an hourly
occurrence, and this will greatly help one to remember what
kind of stock is being passed through every number employed
throughout the system :—

NUMBER OF SILK WITH THEIR EQUIVALENTS IN GRIT GAUZE AND WIRE RESPECTIVELY.

No. 18 centrifugal equals No. 0000	No. 80 centrifugal equals No. 7
,, 24 ,, ,, ,, 000	,, 86 ,, ,, ,, 8
,, 82 ,, ,, ,, . 00	,, 98 ,, ,, ,, 9
,, 40 ,, ,, ,, 0	,, 112 ,, ,, ,, 10
,, 48 ,, ,, ,, 1	,, 116 ,, ,, ,, 11
,, 50 ,, ,, ,, 2	,, 124 ,, ,, ,, 12
,, 54 ,, ,, ,, 8	,, 182 ,, ,, ,, 18
,, 58 ,, ,, ,, 4	,, 186 ,, ,, ,, 14
,, 60 ,, ,, ,, 5	,, 144 ,, ,, ,, 15
,, 68 ,, ,, ,, 6	,, 156 ,, ,, ,, 16

As a further incentive to familiarity with all appertain-
ing to the clothing of machines, another table is added,
viz. :—

Silk,	Numbers	0000	000	00	0	1	2	8	4	5	6
Grit Gauze,	,,	18	24	80	86	44	50	56	60		
Wire.	,.	18	24	80	86	42	48	54	60	65	70
Silk,	Numbers	7	8	9	10	11	12	18	14	15	
Grit Gauze,	,,										
Wire,	,,	80	90	100	105	110	115	125	185	145	

A well-fitted silk cover is of the highest importance, and
has a direct influence on the work of the machine ; and a
few instructions as to the measurement and description of
such, when new ones are required either to replace old and

worn out numbers or to experiment with others not tried before. Measure the circumference of the machine with a tape measure at the head, middle, and tail end of the sheets required. If to be stitched together say exactly what length for each number wanted. In all cases give the dimensions without regard to stretching, because this is allowed for in making up, and check these dimensions carefully. Specify where eyeletting, plain sewing, plain ticking, or corded band is required, and point out where tacking on to the machine's framework occurs. On machines containing division rings it is also best to give both measurements, and width of same, and if a sketch of the cover required can be provided it is an additional safeguard against mistakes.

The best silk come to us from Switzerland. It is woven as a rule 40 ins. wide, and this measure in centrifugal parlance is a sheet. Thus a one-sheet machine will be 40 ins., a two-sheet 80 ins., and a three-sheet 120 ins. in length, inside measurement, or thereabouts. Silk is made of various degrees of strength, such as single and double twist, extra heavy and even of quadruple strength, to suit the several kinds of material being dressed. For instance, the heaviest and best is used for doing work on semolina, which latter, owing to its nature, is free cutting and soon wears away the strands of the lighter sorts ; while softer stock requires, as a rule, finer numbers wherein there is not that absolute necessity for strong threads pure and simple, and besides the nearness of one thread to its neighbour does not admit of its being carried too far in this direction. Silk sometimes varies even in the best brands, and no very satisfactory reason has yet been given why this should be so. Some experts give the water used in its manufacture as the cause, and undoubtedly the water has something to do with it always; in fact, it is given as a principal axiom against its being manufactured in our country that the

constituents of the water, do not favour the best results
being obtained, and the writer has also been informed that
should the season be unfavourable for the proper growth of
the mulberry trees, on which the silk worms are cultivated,
the quality will suffer in that way. Add to these a few of
the reasons why a miller's flour is not always exactly as it
ought to be, it should not be difficult to allow a little
latitude for what might take place in regard to it.

Regarding another prominent machine of the plansifter
class, the author has quite recently been in Germany to
inspect its practical value, and there is no doubt a time of
great usefulness in front of it. Taking the Luther pattern,
it may be said that the machine is built round a solid iron
body, centrally situated. The upright driving shaft is here
also, and on this are two large balance pulleys to prevent
all semblance of vibration. There is a massive wooden
framework to support the sieves, and the latter swing upon
fourfold cane hangers. The central shaft hangs upon ball
bearings. One sight-feed lubricator oils the whole of the
bearings automatically. If preferred, the whole framework
can be had in iron, but Messrs. Luther say that wood, being
more resilient, answers better. By a simple contrivance
the sieves are pressed both inward and downward beyond
the possibility of working loose.

A series of perforated diagonal plates run down one side
of each sieve portion, and these serve the purpose of distributing the material equally over the whole surface, and also
cause the repetition of the numberless little revolutions of
the stock, which is a principle always recognised as essential on this class of flour dresser.

Under every silk sieve is an ingenious brush, which is
ever moving forward about an inch per revolution of the
sifter, and when it has reached one end it begins to travel
back again. This brush cannot simply slide, because there
are small steel pegs which prevent it going too far. At each
end there is a peg left out, and this causes a corresponding

peg on the brush to reverse the movement, and so it goes back again and forward again without any intermission.

The sifter can be driven from either above or below the floor, the total height is about 6 ft.; and if driven from

Luther's Plansifter.

below 6 ins., less will suffice. This is for a 9-sieve machine. The machine can be fed at all corners.

The writer was assured that it was impossible for one of these machines to choke up, but pushing his small prejudices a little further was told that in that improbable event a simple diversion of the feed for 20 seconds would allow all the sieves to clear themselves. All the sieves are simply braced together.

The total weight of the machine is about 82 cwt. There are only two principal sizes built, but either of these can be made to do duty for any size of mill, and it is mostly a question of preference.

The silk on the sieves will last as long as silk on centrifugals.

Plansifters run without any noise, and are very gentle and yet thorough in their action on all the various particles. In shape, in build, and in the matter of brushes the older type is altogether superseded, and it seems now to be under perfect control.

The points most affecting this part of the gradual reduction process may be summed up as under :—

1. Reels, sieves, and centrifugals are employed in the process.

2. All of these to be kept free from clogging in the mesh and speeds to be maintained.

8. Centrifugal beaters to be set at a proper angle, according to the quality of the feed going to it—not too acute to hurry unfinished or dusty material out at the tail-end or cut-off, and not too straight to keep it too long under operation and so tend to drive impurities through the silk.

4. Covers of dressers to be fine enough for local trade requirements, free from unnecessary, or too large, patches, and renewed before the latter are numerous enough to interfere with the machines' efficiency as regards capacity.

5. Speeds to be carefully maintained, otherwise the danger exists of the feed accumulating inside the machine, and, if this be a centrifugal, there will be a liability of the beaters to burst the cover.

6. All parts of machines should be made to work easily by being sufficiently lubricated, especially the inner

revolving cylinder at each end of centrifugals, which latter should be examined at reasonable periods to ensure their not sticking and causing a loss in the use of driving power.

7. Belts to be kept pliable by a slight dressing of castor oil, and laced evenly and smoothly to prevent vibration when running over pulleys, particularly so if the drive is at a twist.

8. Cut-offs and tailings to be frequently inspected, so that the inconvenience of treating dusty products on rolls and purifiers may not prevail to any extent.

9. Flour from all parts of machines needs looking at constantly to note if any breakage exists in the silks or any deterioration in colour or texture arising from some previous operation or machine.

10. Centrifugals should have good holding back gear, in the shape of sprocket-wheels at the tail end of the casing, and the tension pulley, which also acts as a guide pulley, should not keep the chain too tight.

11. The silk covering of all dressing machines should fit tightly; otherwise sieves will sag and accumulate stock, and the silk will probably be caught by beaters.

12. Divisions in machines should be made absolutely proof from leakage by tacking pieces of felt or sheepskin round the revolving edge close to the division boards.

13. Double worms are preferable to enable the operator to classify flour the better by cutting off a few inches at the leaving end or for other and various purposes.

14. Brushes are desirable on all flour-dressing machines, and failing them a piece of soft material such as felt may be run along the roof of the casing, so that the reel just touches it and keeps clear.

15. Avoid having stale flour clinging to the sloping sides of machines by frequent cleaning, run them clean empty now and again, and either send the clearings for lower grades or mix them off again through a mixer.

16. Contrive that covers shall be so arranged that when the rolled material has got over or passed through them neither too little nor too much has been extracted, as judged by the flour or bye-product.

17. Get familiar with all silk dressing numbers and the sort of feed or product they will and will not pass.

18. Look out for little lapses from the strict line of duty of all machines and parts, oil well, keep clean, and never take anything for granted which requires attention.

Millers may be sometimes called upon to make out a list of machines required to deal with a given quantity of wheat or to make a certain number of sacks of flour per hour, and in such instances the following may be accepted as pretty near the mark for a plant making 10 to 12 sacks. It has no reference to the wheat-cleaning department, but refers solely to the manufacture of the finished article from cleaned wheat—viz., 5 four-roller mills 40 inch by 10 inch, grooved for breaks, speed 250 and 100 revolutions; 4 rotary or other sieve scalpers, double, speed 200 revolutions; 4 sorters for scalper over-tails; 1 centrifugal bran duster, 3 sheets, 3 ft. diameter, 24 wire, 250 revolutions; 1 centrifugal bran-meal duster, 3 sheets by 2 ft. diameter, cover 11-12, 16 revolutions; 2 inter-elevator chop scalpers, 3 sheets by 2 ft. 8 ins., at 32 revolutions, 2 sheets of 60 G.G. and 1 sheet of 50 G.G.; 2 chop centrifugals, $2\frac{1}{2}$ sheets by 2 ft. at 200 revolutions, cover 11-12, 6; 16 purifiers, all single, 96 ins. by 16 ins., silk surface covers according to feed, speed of 450 revolutions, fans various; 10 four-roller mills, each 30 ins. by 9 ins., smooth, 250 revolutions for fast roll, 200 revolutions for slow roll; 14 flour-dressing machines, each $2\frac{1}{2}$ sheets by 2 ft., running at 200 revolutions, and 20 revolutions for cylinders of same, covers to suit wheat and flours required; 2 separating eccentric sifters, crank speed 500; offal provider, eccentric sifters, crank speed 500; exhaust fan for rolls and sorters at 600 revolutions, and dust collector; horse-power required 130.

Or yet again, a simple way of expressing the same may be taken as under for one of about the same size, slightly different in detail, viz. :—

BREAKS.

1st—80 inches, fluted 8 and 10 to the inch.

2nd—80	,,	,,	12	,,	,,
3rd—80	,,	,,	16	,,	,,
4th—80	,,	,,	22	,,	,,
5th—80	,,	,,	26	,,	,,

REDUCTION.

A—40 inches—smooth rolls.			G—40 inches—smooth rolls.		
B—60	,,	,,	H—60	,,	,,
C—40	,,	,,	I—40	,,	,,
D—40	,,	,,	J—40	,,	,,
E—40	,,	,,	K—40	,,	,,
F—40	,,	,,			

This plant would therefore consist of :—

9 double 40-inch roller mills
2 ,, 30-inch ,, ,,

A few questions bearing upon the subject matter contained in this chapter are calculated to assist the memory in retaining the fundamental principles involved therein :—

1. Name the chief parts of a round reel and a centrifugal.

2. State the difference existing in the way of dressing on the machines just enumerated.

3. What action governs the employment of beaters? What is their function, and how are they arranged?

4. Give the approximate speeds of beaters and cylinders, and state why the latter revolve.

5. Why should dressing machines have double worms, cut-offs, or holding back gear? What governs the latter device.

6. Make a list of silk, gries gauze, and wire numbers to correspond one with the other.

7. Give speeds of centrifugals, beaters and reels.

8. State the means whereby the stock fed on to a sieve is propelled towards the opposite or tail-end.

9. What manner of cleaning device is employed in sieves, and how is the return or continuous duty performed?

10. Give a list of capacities of one, two, and three-sheet centrifugals with ordinary diameters.

11. What is the reason of dressing through finer silk numbers towards the latter end of centrifugals?

12. For what purpose are sieves used as dressing machines in this department of flour milling?

13. Is it advisable to run centrifugal beaters faster than ordinarily upon any kind of feed?

14. Draw a cross section of the head of a centrifugal showing beaters, reel, and also the lifters described in this chapter.

15. What are the advantages or disadvantages of round 6, 8, or 10-sided dressing machines?

16. Draw a sieve suitable for extracting germ; give revolutions. Define the kind of motion, and state the G.G. number it would be clothed with to ensure a pure product.

17. Define flour; state through what silk number it is generally called dunst. What are middlings, tailings and offal, and why so called?

18. How many times is it usual to reduce the following before they are finished semolina, tailings, dunst and middlings? Do they change their names after one, two, or three reductions? If so, trace them all into the sacking off floor.

CHAPTER X.

FLOUR AND FLOUR TESTS.

What the miller wishes most to know after all the various operations have terminated is : Will the resulting flour be all that is desired ?   And to enable practical men to form more or less accurate opinions of various combinations or mixtures, several modes of procedure have been invented. Wheats, as we have seen in the last chapter, are noted or depended upon for various properties, but at the best our knowledge is but approximate, and more or less of a guess, when it comes to the 'crucial point.   If this is so when speaking of one variety, what must it mean when practical men are called upon to judge beforehand the results of combining or mixing, say, six to ten different kinds. Millers of past generations judged their flour as we do wheat, namely, by handling and by sight, and when these two methods were perhaps the only ones known, they were reliable enough for the purpose, the hand and the eye acquiring, by long practice, a high standard of education—as then existing. This standard was brightness and liveliness.   It did not matter much that the flour had a few specks in it, and, in fact, a few specks do not injure the loaf if everything else is all right.   On the other hand, an attempt was made to prove in the Reduction chapter that spotless flour may very easily be injured in various ways, and tending to produce a

loaf of at least indifferent build, structure, and colour, even
if it did not end in something worse.

The usual plan was to take a handful of flour, shake it
up lightly in the hand, and hold it to the light. If it is
good it will separate, forming a nice lively broken mass,
with a yellow tint in the shadows and an apparent trans-
parency in the particles.   If handled in the sack it is lively,
and works freely through the hand, and if tested for weight
will be heavy in proportion to its bulk.   Indeed, weight is
almost a certain test of quality.   To compare flour by this
method of handling, it is necessary to take a portion in
each hand and loosen as before ; expose both hands
simultaneously to the light—a northern light always by
preference, as it is more subdued—and the first impression
will almost always be the correct one.   It may be that the
light is not reflected equally on both samples, and it is well to
either cross the hands to bring them into opposite
positions, or to change the samples from one hand to
the other ; but it is a peculiarity of this method that the
longer you look the less decided you are as to which is
superior, that is, of course, when the difference is slight.

Taking the example just given to apply to a good sample,
the opposite will be found to obtain should the same test be
applied to flour of inferior quality. When held in the
hand there is not that readiness to fall apart which was
noticed before ; it holds together with a more or less
strong inclination, and when made to part does so in
lumps.   Looked at as a whole, the general tint will lack
the slightly yellow lustre we associate with the best
sorts—it may perhaps be more of a dead white, with a
dark shadow in the spaces where the small lumps are
divided.   Should there have been something left in the
wheat—smut, for instance—the smell will find its way
here too, and depress its character still more.   A dead
white, or chalky white, flour is not to be relied upon,
and should it have a bluish tinge, such flour is worst of

all, and although the loaf baked from it would be white, perhaps, it would lack lustre, be sad, insipid, and the crust would be hard and flinty.

What is known as the Pekar test is about the commonest mode of procedure when seeking to get an idea of the true colour of finished flour, and consists in simply pressing a sample, or several samples, of flour on a small board and then immersing in water. Pekar used small boards with black surfaces, and a piece of plate glass with bevelled edge for the purpose of squaring the edges of the samples so that they would fit closely together, and for pressing them down so as to get a smooth surface. The nearest approach to this I had seen previous to its introduction was the dropping of water over the smooth surface of a sample of flour, which certainly was an improvement over the old mouth-wetting process.

To get a proper test by this method it is necessary to have the different samples on the board of equal thickness and space. If they are too thin the board shows through, and if of unequal size the eye will be attracted to one in particular. There are three ways of examining the flour when using this test. One while the samples are dry, which may or may not be conclusive; another directly after dipping in the water, which will show up any striking peculiarities of colour; and the other after the wetted samples have dried. It is very interesting to notice the changes that take place in these three stages. The qualities mentioned at the beginning of this chapter will have their various colours accentuated. The good flour will appear creamy white and die off gradually to a yellow grey, while the poor flour will show dark beside it, and will die off a dark or whitish grey. I think the true colour is best ascertained in the wet state, and that with some kinds of flour the dry sample is likely to be misleading. But in cases where the wheat is contaminated with dirt, the dried sample is likely to give a truer indication of what the

colour of the loaf will be than either the dry or wet one, as
dirt is sometimes in a granular form and is dissolved by
the water, thus combining more thoroughly with the flour
and imparting its hue to it.

It is only right to state that testing flour for either colour
or strength should be made with a knowledge of the wheat
from which the flour is made. A fairly successful way of
testing by comparison is to lay a small portion on a board,
cut and part it, place another sample in the open space
between the parted lot and then join the latter up again
and press with a spatula. If a difference exists it will at
once show up whether lighter, darker, or different in general
hue. This difference will more readily be seen if the
samples are held up level with the eye in a quiet light. In
judging flour the difficulty comes in when we seek to know
by this means what sort of loaf can be had from it.
Unfortunately, perhaps colour is no criterion of quality,
because the poor rough-looking product of a known strong
wheat will turn out bread of better quality than a finely
prepared highly-dressed flour from a white wheat having no
other qualification, and this is where we, as millers, are yet
at fault. Some years ago Mr. Lovibond, of Salisbury,
introduced to the National Association of British and Irish
Millers his tintometer, by means of which we could maintain
not only a standard of colour, but a standard shade of
colour, and those who are acquainted with the variations of
shade, from pure yellow to pure white, which our wheats
possess will appreciate what this means.

Some men still stick to the older method of doughing up
small balls and allowing them to stand until a dry surface
is exposed, and the colour is gauged in that way; and if
these balls or patties be allowed to stay 24 hours in a
suitable atmosphere they will assume a mushroom shape,
and can, to a certain extent, be judged for the tenacity of
the gluten. It is a good practice to take out and examine
the flour from every dresser regularly, and when this is

done, and the sort of roll feed it is made from studied, in
time a man has only to look into the roll hopper to form a
pretty accurate estimate of the leading characteristics of
any grain portion.   These few simple tests are chiefly
concerned with colour, but there is the more important one
of strength. It is not desirable to again describe the nature
and formation of gluten.   Suffice it to say it is that
property which may aptly be called the life and soul of the
bread.   It supplies the lungs of the loaf which the ferment
made gas expands into a spongy mass, and in which state
it is put into the oven and baked, and if the dough was
not baked when this gas had inflated it the effect would be
lost, because it would revert to its original position and
condition as simply a lump of dough.   It is quite true that
it would make a feeble secondary attempt to rise, but its
sweetness would be lost—it would be sour.   Fermented
dough having passed through these stages will ferment
never again; its life has been destroyed; its soul, so to
speak, has departed, and it is for all practical purposes
dead.   Indeed, although it is not within the province of
this book to say so, all forms of animal or vegetable life
may be likened to a ferment or rising sponge which for a
time upholds and feeds that wherein it is encased, and
having passed the millennium, proceeds more or less slowly
to subside, and finally to cease to act altogether.

There is an excellent and simple way of testing flour,
which, moreover, will give a fair indication of baking value;
this is by boiling.   Take a given and accurately weighed or
measured quantity of flour and add a weighed quantity of
water, say four ounces of flour and two ounces of water,
mix it into a stiff dough, tie it tightly in a cloth, and boil
until cooked, say about three-quarters to one hour, but the
exact time can be ascertained by practice, and varies in
different flours.   Do not over-cook it, as it then becomes
unnaturally hard.   When turned out of the cloth the exact
quality will be seen, especially if several different samples

are boiled together.   If the flour is strong it will swell and come out very stiff and plump.   If weak it will come out wet, relaxed, and sticky.   Moreover, the colours of the different flours will be most distinctly revealed, so it is seen that by this test we ascertain both strength and colour.   When cold, they can be cut somewhat as a loaf might be.

Going back for a moment to the dough-balls, it is best to put a little flour in the hand (which must be clean) and drop into it a small quantity of water, work it into a moderately stiff mass with the fingers, and test its tenacity by stretching it out considerably.   If it works up nice and dry, and licks up the water readily, it is a good sign that it will yield well in the baker's hand or oven, and the probability points to its being honest and sound.   In contradistinction to this, it may work up damp, it may require constant dipping in dry flour to keep it from sticking to the fingers, and then, in the majority of cases, the experienced miller will say it is weak or in some other way unsatisfactory, and that the baker will in all probability find a fault in it.   To explain this test in detail, let us see exactly how it is done.   Work the different samples into balls like large marbles and lay them on a marked paper or board, so as to distinguish the samples, as it is very awkward, after having made tests, not to be able to distinguish them.   Let them lay for half-an-hour, by which time they will be in condition.   Test them now by drawing apart.   The sound, dry flour will pull stiffly and break with a snap.   If it is pulled in a thick mass for a short distance only it will visibly spring back like indiarubber.   If made into a marble and pressed on the top it will almost recover its shape again.   By working it in the fingers, perhaps with a trifle more moisture, it can be drawn out to a considerable length, but it will always break short when it parts.   It can also be spread out into a thin film or blown into small bladders.   In fact, good, sound glutinous flour can be made to do anything.

Bakers' tests are considered generally very satisfactory, and it is an almost universal custom for small, medium, and even for some large mills to have samples sent to be baked whenever a radical change in the wheat mixture is being contemplated. The author has his doubts upon the absolute reliability of these tests, for the simple reason that, taken as a class, journeymen bakers are not very proficient in treating small quantities, also the rule is to take the sponge from a larger batch in preference to treating the given quantity severely alone, because this saves trouble and also, very probably, gives the small sample a lift in the oven, and tends to make it show up better than it otherwise would, so that this in reality is not quite fair, and the test is, to some extent, misleading.

But now for another quality. If, after laying for half-an-hour, the dough appears damp, sticks to the fingers, and when drawn out stretches into long strings, which, instead of parting with a snap, seem to fall asunder as if held by fibres, be sure you have got something dangerous. To further test both samples, or any number, make them into marbles again, and lay them as before on the board, and leave them for several hours—say, all night. A sound flour will maintain its form, and when lifted show scarcely a mark upon the board where it has lain. A poor flour, such as we have tested, will have settled down, and when it is about to be lifted will be found sticking to the board, and will also stick to the fingers most tenaciously. Before doing anything with this kind of flour it should be tested by baking, as that is, after all, the most absolute test; it is most dangerous, as the gluten is either damaged by sprouted grains or from immature wheat. It will be found most frequently in Russian and, of course, in English wheat, but all the causes and conditions are beyond the province of this treatise, although we might probably include poor, worn-out soil amongst them.

There does not seem to be any new test for sweetness; the old fashion of smelling is to us the nearest guide. Flour which is tainted to the sense of smell, or which has a fusty odour about it is a bad sign—perhaps the worst sign—and there is trouble ahead should it be allowed to go into consumption. Really there is no actual remedy for it, and the only way is to mix it slowly—very slowly—with other kinds which are above suspicion. We have read that a peculiar weed or seed will impart an unpleasant odour to flour; it is quite harmless, but at the same time objectionable. Smutty wheat, except very carefully washed, will also lend its baneful influence both to the smell and also to the colour, although it exerts its greatest power in the offal sack. Musty flour will smell the strongest after being fermented, and if very pronounced the fumes from the oven will speak of its presence, and the bread may also be somewhat impregnated.

There is another kind of flour, indeed many, which come in between these extremes, and that is starchy flour. Now, as gluten is absent there can be no elasticity in it. Starch alone works simply like putty, and flour of that kind can be made available for bread making only by mixing with good, strong, glutinous flour, although good dry starch is a most useful antidote to that faulty flour which we have just tested, as it retards the hydrolising or wetting action of the soluble gluten which is present in excess. The degree of gluten can be ascertained by noting the tenacity or otherwise of the flour. All flour possesses some gluten, but some has very little, and practice is required to test and ascertain the proportion present. The same tests can be made as before, but it will be found that if there is a deficiency of gluten the dough will maintain its shape if left for a time, but it will become dry more quickly than the other, and will break apart when tested after laying for a few hours. Indeed the character of starchy wheat can be soon ascertained and determined, and its honesty is unmistakable; but poor gluten is a most dangerous and insidious thing to deal with.

It must be laid down, however, that while the foregoing mill tests are useful and also interesting, nothing is absolutely reliable which does not end in baking.   This is the ideal that the miller aims at, namely, that when everything has been done in the mill that can be done, the baker or customer will find no fault.   An excellent plan is to put every change of mixture through the final or baking test, which, being made a practice of, will save an infinity of trouble in milling what are generally unknown quantities. A few sacks could be ground and a sample loaf obtained from the mixed bulk ; and, in the majority of cases, none of the flour should be delivered away from the mill until this or some other and perhaps more convenient method ensured a knowledge of its quality.   Large mills known to the writer select a sack of each new departure or experiment, and bake it on the premises, and the bread is distributed to the men at cost price. Some flours are noted for their good colour, others for flavour, while again there must be flour in the blend to give stability, body, or foundation to the loaf.   Just as teas are blended to obtain certain results in strength, flavour and aroma, so is it necessary that the resulting flour from a selection of wheats should produce a loaf of bread which shall combine good qualities in colour, lightness, flavour and digestibility.   Summed up, what we expect to find in the finished article is colour, flavour, yield, and finally the all-round quality as calculated to please the consuming public.

CHAPTER XI.

THE WHEAT BERRY.

A grain of wheat is a component body more elongated than round in form; in fact, it is often oval, and is divided longitudinally or lengthwise by an indentation into two equal parts and similar lobes. In colour it varies, but most frequently it is white, yellow, brown, red, or greyish. Strictly speaking, it is composed of three distinct parts, namely, the bran, the kernel, and the germ, and the proportion of each from an average weight standpoint are to the whole as follows :—Bran, 14·4 per cent.; kernel 84 per cent.; and germ, 1·6 per cent. The bran is generally admitted to be about 0·004 inch in thickness, and comprises the pericarp and the external coverings of the kernel. The pericarp again is divided into three parts, namely, 1, epicarp or outer layer, which is very thin; 2, mesocarp or middle layer, formed of elongated cells lying in two or more rows above each other, and running parallel to the crease or indentation just mentioned; and 3, the endocarp, the inner layer, formed of a single row of cells ranged perpendicularly to the direction of the cells of the mesocarp.

The germ is 1·6 per cent. of the wheat (about). Its colour is a clear and very pronounced yellow. In the germ also principally is found that fragrant nutty odour peculiar to this cereal. The germ consists of two parts: (1) the radicule; (2) the gemmule, which is surrounded by four folioles. These two parts form one cellular mass, the cells of which are very similar to those of the seminal integument, to which, indeed, the germ is found to be directly attached. Besides this, it should be added that outside of the epicarp and within the folds of the indentation dividing the lobes of the grain is found the peduncle, by which the grain is attached to the ear. At the opposite end of the grain from the germ is found some exceedingly fine bristles, forming a minute delicate down or beard on the grain. Putting it as clearly as possible, the analysis of the structure of the wheat berry may be said to be expressed by the following table, viz. :—

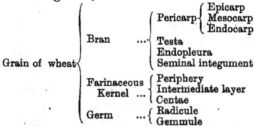

It is only right to say at the outset that these distinctions are very difficult to indicate in well-harvested, sound, dry grains, as by the shrinking and drying after being cut and stacked or threshed, the general structure is somewhat modified, especially in the wrappings or coats of bran alluded to. For instance, we have noted that the pericarp is made up of three minor parts, but they, along with the testa and the two other components, making up the five outer skins, are so absorbed, little by little, into each other that after a time it is very difficult to make out more than two of the original number.

An analysis of the germ may be given as under :—

Starch...	41·22
Albuminoids ...	22·65
Gum and Sugar	9·72
Fat or Oil	5·40
Cellulose	5·97
Ash	3·99
Water ...	11·05
	100·00

Stated shortly, the germ may be said to be the life of the future wheat berry, and consists of a mass of cells full of fatty matter, ready to start into life on being given very little provocation. Forty degrees of heat will cause germination if attended with moisture, and 110 degrees will arrest it in its incipiency.

Analysis of starch :—

Carbon	...	...	...	...	44·44
Oxygen	...	...	...	...	49·89
Hydrogen ...	...	...	...	...	6·17
					100·00

Crude gluten is made up of

Fibrin	...	...	...	...	71·00
Glutin	...	...	...	...	29·00
					100·00

The latter is soluble in water after long immersion, but it is to the former that millers look when testing for accurate and reliable results. Fibrin is insoluble in the time occupied in baking, and it is this substance which holds the gases generated by fermentation and so causes the loaf to be well expanded during the process. The fibrin of gluten also contains nitrogen, sulphur, and phosphorus, none of which are found in starch. These short tables are given for the milling student to remember, and to enable him to more easily understand the references to be made in connection with them at a later stage.

In dealing with wheats it is usually the custom to look at them from a point of view characteristic of some property which the sample under observation is more noted for. For instance, English varieties are relied upon for colour and flavour more than for strength, and in like manner many kinds of Russian samples for strength, without special regard to other qualities which they may or may not possess to any extent, not so prominent; while again Americans can be obtained which unite all the best components in a remarkable degree perhaps not enjoyed by any other pure variety. These given examples are not by any means to be taken in a literal sense, because there are times when the second-named herein fairly surpasses its follower in the main attributes, while English is not always as bad in character as during the last two seasons (1902-1903). But, however that may be, the point to be here noted is this—that the two principal essentials are strength and colour, and the first named is the most rare, being nearly confined to the North-western provinces of America, to Canada, and to Russia, while coloury wheats are fairly abundant in all the wheat-growing countries of the world. This definition is only intended for wheats which are sent to England and used freely in English mills. Hungarian wheat, for instance, is both rich in colour and is also very strong in the loaf raising department, but its use is not so general as to cause it to be a factor in compiling all-round statements. The wheats of France, Germany, Italy, Roumania, Bulgaria, or any other of the Principalities or Powers sending small shipments at fitful intervals, are also not reckoned in a general way, but those who are counted upon each year to supply us in a continuous stream and in somewhat regular fashion with the sorts we from time to time require, and whose produce it is incumbent on all milling students to attempt to gauge at its correct value.

The origin of wheat, as we find it, is wrapped in profound mystery. Whether it has always been, as now, a perfect fruit, or is evolved from some other cereal or seed, we cannot tell. Scientists have lately told us that it is almost impossible to inoculate it; that is, cross it with any seed foreign to it, because directly the pollen or fructifying powder has done its work, the medium of conveyance withers and the receptacle closes firmly up, and the pistolate and staminate are so close together as to prevent this inoculation being performed in any but a natural manner. On the other hand there are not wanting clever men who claim to have discovered a method to circumvent nature and to be able to cross the pollen of, say, wheat with barley, or *vice versa*.   Strictly speaking, however, this does not so much concern the miller as yet.   He is content to learn a little of present practice and leave to future historians to chronicle their own discoveries in their own age.   Wheat, then, we may take for granted, does not inoculate, but is self-fertilising, and we do not learn of any other plant being under precisely such conditions ; and this, doubtless, explains why it is that the wheat berry has never been found in a wild state, and the chances are that an all-wise Creator made this one perfect food when all things were made.   Wheat will also preserve its fecundity or germinating power for a long period away from heat and moisture.   When planted in the ground the germ is the first portion to begin a new life, and it gradually feeds upon the other portions until it has attained sufficient strength to absorb wholly from the soil, and then begin a series of most interesting stages before it stands in all its golden majesty awaiting the common doom.   First comes germination, as just mentioned, next foliation, fertilisation, and maturation.   The second stage has to do with the embryo plant when it is putting out foliage, fertilisation has special reference to a previous remark about the propagation of the species, and maturation refers to the well-developed plant as a whole, and to botanists the culture of wheat is a very attractive subject.   All millers are supposed to know something of the constituents of the wheat berry, that is, as to why there are some for which a decided preference is shown, as against other varieties of seemingly as good quality to an inexperienced eye.

To millers of limited experience it is perhaps just necessary to say that while all wheats come under the broad definition of constituents enumerated, the possession of those which are most essential from a miller's point of view varies in a degree nothing short of being remarkable. It is, no doubt, known to the youngest member of the

trade that England does not grow more than a fraction' of
the amount of wheat necessary to feed its population.
What is perhaps not so well understood is the fact that
should the quantity be largely increased British millers
would still be obliged to use a large percentage of wheat
grown abroad, and it may naturally be asked why this is so.
All wheats contain flour, it is true, and if we look back at
the composition of them we must admit that all wheats
contain the elements for making bread. Every berry is a
perfect fruit in itself, so why should not one be as good as
another in every respect? There is just one vital reason
lying at the bottom of the answer, and it is this—the
quality of the gluten. Experience teaches us that it is
impossible to obtain this from English varieties, and for
various reasons. To start with, it is not in native sorts to
absorb this nitrogenous matter beyond a certain degree;
secondly, the nature of the soil does not contain it in its
highest form; and thirdly, if the soil contained it in the
quality and quantity required the climate is not conducive
to bringing it to a high state of maturity.

Those millers who have read my friend Mr. Voller's book
will find these statements verified by a series of experiments at home and abroad, which leave no doubt as to the
soundness of the theory advocated. English wheat cultivated abroad loses little of its original character owing to
its nature, and foreign wheat deteriorates on English soil to
a considerable degree. Originally weak wheats do not get
strong on stronger ground although strong wheats lose
their character on weak soil, so that it would appear as if
all wheat-growing countries cultivate the brands most
suitable to their surrounding, both above and below the
ground level. This shortly explains a part of the subject
under notice, and, in order to make it fully understood, we
will see what the gluten is expected to do when present
in sufficient quantity and quality in the flour as manufactured according to the methods set forth in the previous

pages. Gluten, or insoluble albuminoids, is absolutely necessary to the baker. When a sack of flour is about to be baked into bread it is first put to sponge—that is, it is mixed with a recognised proper amount of yeast and water—generally 1¼ to 2 lbs. of the former and 60 to 70 quarts of the latter and from 8 to 8¼ lbs. of salt. This is allowed to stand 6 to 7 hours in order to ferment or sponge. It should be stated that not all the flour or ingredients are used for this first stage, but a portion only, the rest being added at the expiration of the period named. After the whole is leavened it is cut, weighed, moulded and put into the oven to bake, and now the crucial test is being made. Should the original flour not contain gluten of the right quality the heat which causes the loaf to rise under the baking process will perforate the glue-like substance, which up till now has consisted of an innumerable quantity of small hollows or cells encased by this matter, and the bursting of these cells lets the expanding dough fall down again, resulting in a more or less puny flat loaf, and proving that the strength of the gluten was not equal to the task of withstanding the heat pressure necessary to the proper performance of the operation.

On the other hand, should the matter named gluten succeed in holding the generated gases for the whole period of 50, 60, or 70 minutes, its strength is considered proved, and the size, shape, and texture of the bread will be satisfactory at all points connected therewith. It should, perhaps, be stated that the baking oven is heated to something like 500 degrees Fahr., so that the albuminoids are coagulated, the starch is gelatinised, moisture is driven off, and the bread obtains the desired flavour through the various chemical changes it is forced to undergo in the process, providing all the other constituents are in a perfect condition.

Following upon this it will be instructive to glance at a

few tables showing the sources of supply from abroad, with particulars of their various characteristics and values, and for them to be the more interesting they are divided and classed in several ways. A study of them will benefit those mainly concerned, and may be read with profit by all. A few sorts are selected from those of English growth, and are fairly representative of the whole.

VARIETIES OF ENGLISH WHEAT.

Nursery.—A red wheat, generally the highest priced of all native sorts, heavy, yields well, and the strongest variety grown in England.

Red Lammas.—This wheat varies according to soil, but is one of the best old sorts, hardy, good flour yield, has red straw, red wheat ears, and is a red berry, thin skinned and of good gravity.

Talavera Red is also a favourite in most places where it abounds, does not come up to the two former sorts in the matter of flour yield, but matures early and, as English wheats are calculated, is reckoned fairly strong.

Golden Drop.—Mostly grown in the south, and is a thin skinned, red wheat of high flavour, and when harvested in prime condition mills excellently.

Burwell Red is said to be mostly grown in Northampton-shire, hardy, large, and reddish cast, yields an average, and is palatable.

Browick Red is a rather coarse variety, long, stiff straw, and clean, large, plump grains, useful for filling up purposes.

Rivett's wheat is a well-filled grain, and above the average size, soft in texture, and with gluten of a very weak order, fairly good yielding sort, and is also sweet to the taste.

Chidham is a white and rather heavy variety, mills well,

almost round in build, and remarkably even in size. This wheat is said by experts to be an off-shoot from a Dantzic breed; it is prolific, smooth, ripens in good time, and requires to be cut before it is dead ripe; otherwise it is liable to shed out.

WHEAT TABLE No. 1.

WHERE GROWN.	SORT.	COLOUR.	MONTH OF HARVEST.	FIRST ARRIVALS IN ENGLAND.
Russia, North {	Saxonska ...	Red	Aug., Sept.	Oct., Nov.
	Kubanka ...	Yellow	,, ,,	,, ,,
Russia, South {	Azima ...	Red	July, Aug.	Sept., Oct.
	Ghirka ...	,, ...	,, ,,	,, ,,
	Taganrog ...	Yellow	,, ,,	,, ,,
Germany ... {	Dantzig ...	White and mixed	,, ,,	,, ,,
	Rostock ...	,, ,,	,, ,,	,, ,,
Roumania Bulgaria & Danubian Provinces.	Usually known as Danubian {	Red	,, ,,	,, ,,
		,, ...	,, ,,	,, ,,
		,, ...	,, ,,	,, ,,
	Varna ...	,, ...	,, ,,	,, ,,
Turkey ... {	Bourgas ...	,, ...	,, ,,	,, ,,
Persia ...	—	Mixed	Mar., April	My., June
	Bombay ...	White, red & mixed	Fb. Mr. Ap.	,, ,,
	Calcutta ...	Mixed	,, ,, ,,	,, ,,
India ...	Delhi ...	,, ...	,, ,, ,,	,, ,,
	Kurrachee	White, red & mixed	,, ,, ,,	,, ,,
	Nagpore ...	Red	,, ,, ,,	,, ,,
	Atbara ...	,, ...	,, ,, ,,	,, ,,
Australia ...	S. Australia	White	Dec., Jan.	Apl., May
	Victoria ...	,, ...	,, ,,	,, ,,
New Zealand	—	,, ...	,, ,,	,, ,,
Canada ... {	Manitoban	Red	Jy. Ag. Sp.	Sept., Oct.
	Soft white	White	,, ,, ,,	,, ,,
	Soft mixed	Mixed	,, ,, ,,	,, ,,
	Duluth ...	Red	July, Aug.	,, ,,
United States of America	Spring ...	,, ...	,, ,,	,, ,,
	Red winter	,, ...	June, July	Aug., Sep.
	Kansas ...	,, ...	,, ,,	,, ,,
	Oregon ...	White	,, ,,	Jan., Feb.
	Blue stem	,, ...	,, ,,	,, ,,
	Californian	,, ...	,, ,,	,, ,,
	Walla Walla	,, ...	,, ,,	,, ,,
Chili ...	—	White and mixed	Dec., Jan.	Apl., May
	Rosario ...	Red	,, ,,	Mr., Apl.
Argentine ...	B. Ayres ...	,, ...	,, ,,	,, ,,
	Santa Fé...	,, ...	,, ,,	,, ,,
	Uruguay ...	,, ...	,, ,,	,, ,,
	Saldome ...	Yellow	,, ,,	,, ,,
Egyptian ...	—	Mixed	Mar., April	Apl., May

Hardcastle is also a name for a white variety, having much the same qualities as Chidham.

Hunter's White is chiefly cultivated in Scotland. Very similar to Chidham grains, opaque and white, and some of a reddish cast and hard, thin skin, and good to work.

Essex White or Rough Chaff is a splendid looking wheat, and generally commands a good figure when offered for sale. Mostly grown in Essex and Midland Counties. The grain is well nourished, and the straw is stronger than most varieties. It is liable to shed out when being harvested.

Webb's Challenge is almost as well known as any English sort. The grains are white and of fine quality usually. Of vigorous growth mostly in Lincolnshire and adjacent counties, it is a great favourite when two years old, and yields very well in the percentage of flour.

Hallett's Victoria is also a white berry of a popular order, and makes good flour both in quality and quantity when proper attention has been paid to the soil it is grown on.

———

Russia sends a quantity here regularly, and the names attached to the provinces growing them or the ports shipping them are responsible for wheats varying in analysis to a greater or lesser degree except in one particular, and that is gluten. Coming after America, and even sometimes surpassing that country, a Russian sample will be mainly judged for its strength. Of certain kinds it may also be said that the combined qualities of colour and strength are blended in a remarkable manner; but the broad fact will always remain of their being mostly in demand on account of their glutinous properties. Colour we can usually obtain much more easily, having a larger area to select from; but for loaf raising we are almost confined to America, Canada,

and Russia. Saxonska is a variety well known in England,
and has a high reputation. It is a red berry, rather small in
build, almost transparent, hard in structure, thinnish skin,
·and fairly clean. The impurities present usually are cockle
and other round seeds, oats, and occasionally barley, rye,
stones and garlic. The quality of the gluten is very good,
and on the whole it is, perhaps, the most reliable of all the
large number of brands which hail from the dominions of
the Tsar of All the Russias. Its weight per imperial bushel
will average 63 lbs. to 64 lbs., and it has the further merit
of working well in English mills.

Kubanka.—This wheat is very hard of structure, fair in
size, flinty almost, thin in skin, weighs up well, is a light
red or yellow in colour, very clean, but it does not come to
us usually in sufficient quantity; therefore it cannot be
relied upon as a staple ingredient in any mixture. The
usual impurities found here are seeds and chaff. This is
one of the varieties which may, perhaps, be classed as being
better in all-round character than an average American
wheat. It will undoubtedly yield a greater amount of bread
than its rival, but in some cases the volume of the loaf is
not quite so finished. Weight for weight the Russian of
this kind will beat an average American, and conditions of
trade locally must determine which—if either—it is most
advantageous to employ.

Odessa Ghirka is a somewhat thin, small grain, hard and
brittle when imported, has a good skin, weighs well, mills
nicely, colour light to deep red, and it is more variable in
sample and in the amount of admixture than the
previously-mentioned sorts. The quality of the gluten is
usually high, and the bread a good colour also. Contained
with it is rye, cockle, dead grains, stones, and not infrequently oats.

Taganrog, Nicolaeiff, Marianople, and Berdianski partake
more or less of the same characteristics, and need some-

times to be handled with care ; especially is this so when rye is present in abnormal quantities, because every little extra of this bye-product by so much detracts from the known strength of the pure wheat.

Azoff, Azima, or Winter wheat is a grain of medium size, deep red in colour, with a skin of moderate thickness, and is fairly hard, being sometimes inclined to be brittle. It is considered a strong wheat, containing a good share of insoluble gluten. It weighs well naturally, but usually contains stone, rye, oats, cockle, chaff, pieces of hard soil, dirt, and sometimes smut. Care is needed in the wheat-cleaning department when these latter are present, and if convenience does not allow of its being properly conditioned —washing being essential to bring out its best qualities— it were better left alone, and some other sort depended upon to give strength to the flour.

Such, then, in brief, are the salient properties of the multitudinous varieties known as Russian wheat, and for further matter, both in connection with these and with other wheats which follow, the reader is referred to the accompanying tabulated lists.

American Spring wheats are known and famed perhaps best of all ; the mere mention of them conjures up visions of lofty, well-piled bread of excellent colour, flavour, and strength. The term "Spring" covers Duluth, Minnesota, Milwaukee, and Chicago, and the remarks just applied to them refer to the best or No. 1 variety only. A secondary grade exists which is not quite so reliable, and if used freely the miller probably has something besides upon which he can rely in case these No. 2 sorts should have soft or sprouted grains among them. The wheat is almost transparent, plump, without attaining to undue size, hard to a nice degree, fine, healthy looking, red in the colour of the berry, and to an experienced eye there is nothing more calculated to give one more pleasure than the appearance of a rich sample of this wheat from the great north-western provinces of the United States of America.

American Winter wheat is another variety which produces a very good flour of sweet flavour; it has also a fair amount of strength, and for its all-round qualities is thought very highly of by British and Irish millers. The yield is also fairly good, though the grains themselves show much variation in size. It has a rather thick skin, and the structure is of just that kind which best suits the break rolls, mild without softness and easily deprived of the flour it contains; while the latter has a nice flavour in the bread. The texture also is above the average; and this wheat milled alone will produce a fairly presentable loaf. It is also, as a rule, very clean, and a good sample will run our native sorts very close in this respect. Grass seeds, chaff, some cockle, dust, and occasionally garlic, will be found as admixtures.

Michigan grows a good quality white wheat of medium structure, which has mainly the attributes of winter wheat quoted above in all its particulars.

Californian wheat varies much in the size of the berry. It is a white wheat of a high class value, for colour alone ranking at the very top of the lists perhaps. It is below Australian in other respects, but for pure whiteness or bloom it is indispensable in first-class mills, and by the majority of millers is considered to be unequalled. It is also usually very clean, short straw, barley and oats being the chief impurities which are found. Some years ago there were a few cargoes found to contain a round scented seed called melilot, but little has been found recently. It was a peculiarity of this seed that its elimination did not cure the resulting flour of a spicy taste, and nothing short of washing the wheat was found to be efficacious for this purpose. The bread from this wheat alone is only poor in appearance, rough in the crust, and of a dry coarse nature, but when conditioned and mixed with others it possesses an intrinsic value never for a moment to be lost sight of.

Walla Walla is of small, but pretty regular in size, and is a white variety. For general excellence it may safely be put as a secondary Californian, having in a lesser degree more or less the qualities noted in that particular sort, and is better, according to some millers, when it is two years old.

Oregon is also a white variety of the Californian type, and is of medium structure, not so harsh or brittle, perhaps, and is also very weak or deficient in the gluten department. It also is clean, and oats, chaff, barley, and seeds are, as a rule, easily extracted by modern appliances.

Turning to India, we find there a large assortment of wheat having, in fact, but one thing in common, and that is a more than essential quantity of admixture. Agricultural science is at a lamentably low ebb in that great country, and circumstances do not seem to point in any decided direction towards an amendment. From almost every province the presence of earth balls, stones, and other extraneous matter causes a serious depreciation in intrinsic value on English markets, because the cost of getting the wheat clean is calculated not only upon the machines used for the purpose, but also upon the net loss on the grain itself, seeing that we must sometimes knock off 12 per cent. as a dead loss. This side of the question is always before the miller in purchasing Indian wheat, and the facts just quoted are responsible for keeping them at a lower level than that to which they would otherwise belong. In the Delhi district a fine sample is grown—perhaps the best in the Empire. It is a softish white, well-filled berry, and very uniform in build, and also weighs heavy. No. 1 Calcutta is a mixed red and white variety, and as Indians go, would be called a strong, useful wheat; but the quality varies much according to the season. It lends itself admirably to the washing and conditioning process on account of its ability to absorb a great amount of moisture, being as a rule very dry. Ordinarily the sample consists of about 80 per cent. white and 20 per cent. red, and the impurities of all these brands cover a wide field. It is no unusual thing to find stones, dirt, peas, barley, grain and rape-seed, chaff, weevils, and linseed in one sample, and, as said above, the presence and amount of so many impurities detracts from its real or intrinsic value. No. 1 Bombay is a fine looking large white wheat, soft to flinty in structure, but of fair average density. The skin is medium in thickness, and the flavour of the flour is somewhat starchy or ricey. After treating on a full line of cleaning machines, such as we now use, however, this is entirely got rid of, and, therefore, does not count. Millers who use this as freely as it is available speak of its wonderful pure whiteness in the flour and bread, and there is not much doubt that, properly treated, it has few rivals in that respect. But, as was said just now, these wheats possess properties valuable to the British miller if the latter has the necessary machinery to bring them to perfection, and as ideas permeate more deeply regarding the preparation of wheat for milling, these varieties will command even more attention than now. Kurrachee is a flinty variety, of moderate strength, though the peculiar flavour is present when imported. It is a fairly useful wheat for straight-run flours, as, indeed, are all Indians, but they are not very much good where a high-class strong patent is wanted, except, perhaps, in very small proportions, and that for colour alone in case nothing better for general excellence is to be had.

Hard Fyfe is a grand variety grown in the north-west of Canada. It deservedly ranks in the very front of the world's wheat produce. The bread from a choice sample is superb, lofty, porous, and elastic, leaving nothing to be desired. It is capable of carrying a great proportion of weaker sorts, and can generally be relied upon to impart

great strength to the mixture.   Its impurities consist
chiefly of cockle and round seeds, buckwheat, oats, and
chaff.   The colour is red- and the skin transparent, as,
indeed, are all wheats more or less which claim to possess
the strength element, viz., nitrogen.

Ladoga is also a Canadian wheat, which is usually
harvested early on account of its ripening more quickly
than others grown there.   It is a sort of secondary quality
and is not so reliable in working for gluten. as Hard Fyfe ;
taken altogether, however, it is a very useful kind.

· Northern Manitoban is also a red wheat of full strength
and choice colour, and ranks well up to Duluth for general
excellence.   It has most or all the attributes ascribed to the
latter, and it many times makes an admirable substitute
when Duluth is not to be had.

Australia, perhaps, sends the most handsome grain of
wheat it is possible to find, and the yield of flour is, without
doubt, at the top of the list.   In structure this wheat is
medium, more brittle perhaps than winter American, but
yet somewhat mild like that variety when parting with its
flour.   The skin is of a yellowish or straw-coloured tint and
somewhat thin, of just fair strength.   It imparts a lustrous
tint to the flour which is unapproachable by any other
kind, and the small quantity arriving here always commands full market value.

New Zealand also contributes a quota to English markets,
but, as a rule, these wheats do not rank higher than white
native of medium structure and secondary quality.   The
amount is never likely to upset ruling prices, and they are
chiefly used as neutral quantities which go to cheapen the
ordinary mixture.

Chilian is another white wheat capable of improving
blends which lack high colour.   Of no account for gluten,
it is very useful for filling up purposes.   While being somewhat variable in both condition and structure—some grains

are moderately soft while others are more or less of a brittle
nature—washing and conditioning show up its pure colour
and make it in demand wherever these appliances are to
be found.   Cleaned on a dry system only it has a blueish-
tinted flour, which is often apt to do more harm than
good if used at all freely.   The skin is thin and the yield
good. Impurities, such as chaff, seeds, stones, oats and
barley are usually found to the extent of 5 per cent. to
8 per cent. Baked alone the loaf would be found very low,
showing an absence of glutinous property.

Egypt sends us a small amount of wheat, but agriculture in the land of the Pharaohs is in a very primitive
state, and her cereal produce is on the "bottom rung" of
the ladder of commerce. It possesses no good qualities
and its presence excites no interest ; it has a bad reputation all round. The colour is poor, strength is absent, and
the flavour is unpalatable. It is most notorious for the
large amount of barley and the plentiful supply of mud
present, and its milling value is very nearly nil. In
structure it is medium hard and the skin is brittle and
thick. The colour is white or mixed.

Persian wheat is a mixed sort of inferior quality, and
contains a large amount of impurities. The structure is
hard and brittle ; the colour is fair, but the almost total
absence of glutinous property generally stamps the variety
as useless for even medium class flours.

Taken generally the following are the impurities likely
to be found in a summary of foreign wheats sent to this
country :—

1. Other kinds of grain, such as barley, oats, rye,
   maize, peas, beans, tares and vetches.

2. Heavy round and polygonal seeds, such as cockle,
   charlock, hariff, clover, mustard, and others.

3. Light small refuse, as grass and hemp seeds, chaff,
   smut, screenings, and various mites and insects.

4. Light large refuse, as straw, string, wood, &c.

5. Mineral and soil refuse, as stone, slate, coal, dirt, clay and dust.

6. Metal substances, as nails, screws, tacks, and the like.

The impurities may not all be found in any one sample, but spread over a season all will be discovered in the various cargoes constantly arriving and *en route*.

The chemical constituents of the wheat berry are (1) organic and (2) inorganic. The organic substances are gluten, albumen, cerealine, starch, vegetable fibre, germ, and sugar. The inorganic substances are ashes, salts and water.

From well-conditioned and well-cleaned wheat a flour may be produced having 65 per cent. of starch, 11 per cent. of gluten, 14 per cent. of water, and the remaining 10 per cent. being made up of albumen, cerealine, gelatine, oil, resin, and phosphor. When the wheat first begins to form itself in the ear the berry is built up of a great number or series of cells almost of pure starch. This substance appears as if it is the first filtrate from the food in its original liquid state as supplied through the green straw from the soil and the atmosphere, and is the highly-concentrated gases and solids which eventually become the staple food of man. The process of feeding is ever going on, and it would appear as if the continually-growing substance is pushed outside, undergoing various chemical changes the while, and subject by-and-bye to the outward influence of light and heat, whereby they assume a settled form and proceed to mature, subject to favourable outside surroundings. They reach their proper size and shape at the time the sun heat dries up the source of food—viz., the life cord. They then harden or mature and concentrate their various properties towards a milling condition.

WHEAT TABLE No. II.

Wheat.	Impurities.	Per Cent. of Impurities.
Duluth ...	Cockle, buckwheat, screenings, chaff, dirt ...	2 to 4
Walla Walla ...	Chaff, straw, barley, &c.	2 ,, 6
Chilian ...	Stones, chaff, grass seeds, oats, &c.	5 ,, 8
Californian ...	Straw, chaff, oats, barley, &c.	2 ,, 5
Red Winter ...	Cockle, buckwheat, chaff, dirt, garlic	3 ,, 5
Chicago Spring	,; ,, ,, ,, oats	3 ,, 5
Minnesota ...	,, ,, ,, ,, ...	3 ,, 5
River Plate ...	Stones, dirt, barley, oats, round seeds, &c. ...	4 ,, 10
Bombay ...	Stones, dirt, vetches seeds, barley, chaff, weevils...	5 ,, 12
Kurrachee ...	,, ,, ,, ,, ,, ,, ...	5 ,. 12
Delhi	,, ,, ,, ,, ,, ,, ...	4 ,, 10
Calcutta ...	As above, and peas and linseed	6 ,; 10
Manitoban ...	Cockle, buckwheat, smut, and round seeds ...	2 ,, 4
W. Canadian ...	,, ,, ,, ...	2 ,, 4
Australian ...	Barley, chaff, oats, &c.	2 ., 4
New Zealand ...	Chaff, dirt, smut, &c.	2 ,. 5
Azima ...	Stones, dirt, cockle, garlic, smut, ergot, rye ...	4 ., 8
Ghirka ...	,, ,, ,, ,, ,, ,, ...	4 ,, 8
Polish	,, ,, ,, ,, ,, ,, ...	3 ,. 6
Konigsberg ...	Round seeds, chaff, straw, &c.	3 ,, 5
Saxonska ...	Dirt, stones, chaff, smut, ergot	3 ,. 5
Egyptian ...	Balls of hard mud, stones, loose dirt, seeds, and impurities of all kinds	8 ,. 20

CHAPTER XII.

Protection from Fire.

The establishment of the Fire Insurance Company in connection with the National Association of British and Irish Millers has undoubtedly induced many members of the trade to simplify their methods and minimise the risk of fire; but however that may be, it is necessary that practical, or would-be-practical, millers should be conversant with the older ways of meeting this dreaded foe to flour mills. Before we go on, it will perhaps be as well to give an explanation of the newest plan in connecting what are known as sprinklers with water pressure, to ensure an automatic flow should heat be present from any cause to any extent over 150 deg. Fahr. The latest regulations specify that there must be an abundant supply of water from two independent sources. Usually they are supplied from a tank on the top of some portion of the mill for one supply, while the town main supplies the other. Both these must have a constant pressure sufficient to throw a jet of water at least as high as the mill itself. So much for the water. Now, connected with these two sources of supply are a series of smaller pipes. These latter are arranged in rows, at suitable distances apart, upon the ceiling or roof of every floor. Attached to these are what are technically styled sprinkler heads, and in reality it is upon these little devices that the success of the operation depends. In detailing their construction it may be best to say that the vital parts are held together by a soldered joint, and upon the heat being sufficient in their immediate vicinity to melt this joint, the whole thing drops to pieces and thus releases the resistance before offered to the water pressing behind it. This resistance being overcome, the water rushes out upon a disc or saucer, which deflects its volume and gives it command over some 60 square feet of space. Should the heat continue to extend, as it would if one of these sprinkler heads failed to subdue it, then another one would open and a second spray would commence to operate, and so on according to the seriousness or extensiveness of the heat or fire; and as the two main sources of water supply are supposed to be kept at the constant pressure just noted, they would continue to operate until that pressure was relieved. This then, in short, is the principle upon which these ingenious, clever, and yet simple appliances do their work, and do it well, and the writer recommends their full and complete installation wherever such can be accomplished.

Failing complete protection on the most modern lines as just touched upon, the following means may be recommended as a more or less adequate protection according as mills are situated and the ideas of those in charge coincide, namely :—

1. By having buckets filled with water placed on the floor or hung up at a suitable height and at convenient distances.

2. By having a few hand pumps handy and in working order at well-known places in the mill.

3. By having a series of standpipes connected with the main supply on every floor, and lengths or coils of hose attached or handy thereto.

4. No candles to be used about the mill; covered lamps should always be used if other means of light are not available.

5. The mill staff should be drilled occasionally so as to be able to act promptly should the necessity arise for doing so.

6. On the premises of large mills a fire engine should be kept and all the requisite tackle, and a competent man in charge.

7. Vegetable oil is much safer than mineral and should be used in all covered lights ; and the lamps should not be left about anywhere.

. 8. If possible the mill should be lit throughout with electric light.

9. Communication with fire brigade headquarters should be by wire.

10. General cleanliness inside and outside should be aimed at by all operatives so as to minimise risk.

11. No matches or smoking allowed on the premises.

12. No accumulation of any kind of mill refuse, sweepings, or damaged goods likely to generate heat.

13. Frequent testing and overhauling of whatever constituted the principal safeguard in relation to the question of fire.

There are many kinds of sprinklers on the market, among which are the " Morris " and the " Grinnell." They differ slightly, however, in their manner of mechanism, but all have the same principle, namely, a fusible solder joint to allow of a rush of water whenever the temperature in their immediate vicinity rises above a certain degree.

CHAPTER XIII.

MOTIVE POWER.

It is now-a-days necessary that a practical miller should know something about the origin of the motive power generated to do the work over which he is exercising his ingenuity from day to day. It is not strictly in the line of flour-making proper, but is, perhaps, the most valuable and indispensable adjunct in bringing the work of a mill to the highest perfection. The special point insisted upon throughout the whole process is continuity. This continuity cannot be . practised without a regularity obtains in the motion of the machines or their parts. Everything, in a sense, must be proportional. Every shaft, pulley, belt or other travelling or rotating appliance is calculated upon a certain basis. That basis is speed—peripheral or otherwise. Each separate portion which is included in a mill's complement depends mainly upon doing its correct duty, at a speed calculated beforehand, and unless it does this with unfailing regularity, there is a want of evenness in the general result which is not desirable. This is very well understood by millers themselves; hence we have a state of things greatly different from what it was 15 years ago, and before this point was made as prominent as it is to-day. In the first hurry and rush, this question of motive power was somewhat overlooked, but the writer is glad to testify to the very excellent progress made once its importance became recognised. Shortly, then, we will glance at the various methods which enable a miller at the present time to overcome the resistance of stationary machinery, and so utilise these methods for the bringing about of such results with which our everyday practice brings us into contact.

It is, perhaps, just necessary to glance at the various kinds of motion derived from water. The modes of converting water-power to use in driving mills are many, including undershot wheels, breast wheels, overshot wheels, and the various forms of turbines. The undershot wheel is used upon the lowest falls, which also combine a large volume. The breast wheel and the overshot wheel are pretty much alike, the latter, perhaps, having some advantage on account of its receiving a greater impact from the stream rushing to fill its buckets. The turbine must be reckoned the very best of water motors, because it is free from the cumbersome machinery associated with its more ancient relations. It takes up less room and gives out a greater percentage of the initial force than any other type. For comparison the following table may be taken as fairly accurate :—

Turbines will give 80 per cent. out of the nominal power available.

Overshot wheels give 66 per cent.

Breast wheels give 60 per cent.

The number of turbines is very great, but they are mostly divided into Radial, Axial, and Combined. In explanation of these terms it may be stated that in the Radial the water flows at right angles to the axis. In the Axial the flow of water is parallel to the axis. In the Combined both systems are used.

All turbines belong to one of two systems—viz., reaction and impulsive. The first works with all its parts full, and it is requisite that there be a continuity of flow in every part. The other system includes those which are only partly filled with water, the air having access to the remaining part. Perhaps the most efficient of all is a reaction turbine with a combined flow, viz., inward and downward.

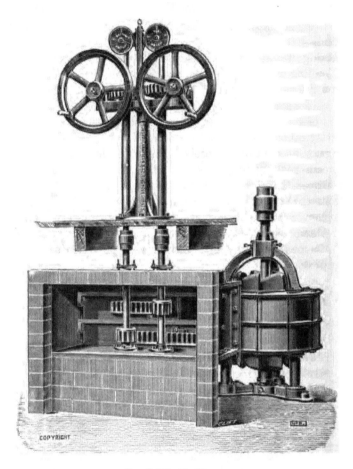

Armfield's Turbine.

Gas engines may be used where small or intermittent power is wanted. No boiler is required, and they are ready for use in a very few minutes; the principle is one of continuous explosions forcing the piston rod backwards and forwards, and after starting requires very little attention.

Steam-engines are without doubt the most useful types of power generators, and may be classed under six heads, viz. :—

1. High-pressure, single cylinder.
2.   „   two cylinders
3.   „   three „
4.   „   single, with condenser.
5.   „   two,   „
6.   „   three,   „

High-pressure engines are considered to be the most economical, provided the cylinder space is so designed as to expand the steam down to nothing but vapour. The effect of the condenser is to add a pressure on the low-pressure cylinder of nearly 15 lbs. per square inch, thus augmenting the power. Condensers, however, are of no use where a proper quantity of cold water is not available; The manner of placing the cylinders vary, but perhaps the favourite one is for them to work side by side, or in very large mills double tandem. Triple-expansion kinds are mostly confined to those giving out at or over 200 horsepower.

The mode of procedure is shortly this :—Steam enters the first cylinder at the pressure indicated on the boiler pressure gauge. It is regulated as to quantity by cut-off valves (Corlis, for instance), which latter are also connected with the governors, and so regulate the number of revolutions. Having done its duty here, it enters the low pressure cylinder for a second exertion, and then to condensor, where, meeting a stream of cold water, or coming into contact with water-charged tubes, it is condensed from vapour to hot water, and may again enter the boiler at a goodly heat, after being relieved of any oily substance contained therein by means of filters, extractors, or the like.

The writer is familiar with many of the latter-day types, viz., horizontal compound condensing engines, and to his mind there is nothing to beat them where correct motion is required.

To judge intelligently of the attributes of turbines, it should be borne in mind that, as generally constructed, they derive power solely from the momentum, or moving force, of the applied water ; and that this force or energy of motion is always represented by the water's velocity. Then, duly noting the self-evident fact that the measure of the power taken from the water is simply the difference between that which it has (as represented by its velocity) at its introduction or application to the running wheel, and that retained in it (as represented by its velocity) at its departure from the wheel, it will be seen that a portion of this abstracted force is unavoidably consumed by friction, &c., which to a greater or less extent is common to all wheels, while in many a much larger portion is wasted by incorrect action, due to imperfections of design and construction.

The author, having had a more than ordinary amount of experience in country mills, has always had in mind the drawbacks under which many millers labour in manufacturing flour to compete in the open market.   These drawbacks are mainly in the power department, and these matters are dwelt upon here with a view of imparting a wider range of knowledge in connection with them.   The most prominent points of superiority which are embodied in turbines are, in the main :—

1. Economical use of water—that is, higher efficiency with any given amount than by any other similar mode.
2. Greater power in much less compass.
3. Steadiness and regularity of motion under all circumstances when fitted with a governor.
4. Ability to move gradually and more under control.

In order that millers generally may know how to measure the quantity of water available to run the mill to the best advantage, it is best to select some spot in the mill stream where there is a moderate current, or a smooth, even flow of water, and measure the depth and width of the stream ; and also measure the velocity of the stream, by timing the passage of a float between two fixed points, the float to be weighted sufficiently so as to sink well into the water.

If the stream be of considerable width, and the flow of water not uniform the entire width, the velocity should be found near the shore, and also in the centre, and the average velocity in feet per minute taken ; and if the bed of the stream be uneven, the depth should be ascertained at a number of points in width, and the average taken.

To obtain the cubic feet, multiply the depth by the width, and this product by the velocity in feet per minute, which will give the number of cubic feet which the stream will discharge in one minute. Then measure accurately the amount of fall that can be obtained.

This data forwarded to any engineer will bring back accurate information as to horse-power available.

The key-note of all the different modes of driving is economy—economy in space, in use, and in application, and to still further benefit consumers, millers included, we are constantly meeting with new attempts in this as in other departments. Most readers are familiar with some kind of feed water heater, a contrivance calculated to heat the water previous to its entering the boiler, and so save fuel, but what is of the most importance in connection is the few attempts, and the fewer successes, which attend the efforts of those who try to utilise more of the heat generated in combustion.

It is here that the total horse-power exerted to run any combination of machinery is determined, and a short description of how to do it may not be amiss, owing to its importance in determining the total cost of production. It is well known that one horse-power is the exertion necessary to raise 33,000 lbs. weight one foot high per minute.

The amount being given out by any engine is calculated by an indicator apparatus fixed to special cocks on the high and low pressure cylinders severally, and which mark a paper covered brass tube with the information wanted, both as to power given out and also as to regularity of steam supply, loss or deficiency of same, or any other defect connected with the mechanism of prime motion. To calculate the cost in fuel is an easy matter once the total horse-power used is known; we have but to total the amount of coal or other substance used, at the price per ton or other weight, the number of hours run, and the amount of flour made in sacks, and the answer to this question is a matter of very simple arithmetic, which can be used to determine both for horse-power indicated, coal consumed, and cost per sack of flour per hour. It is a wise plan to put this test into operation once every three months to make sure that nothing is being wasted.

To find diameter of cylinder for a given power : Multiply horse-power of engine by 33,000. Divide product by the product of the cylinder area, plus steam pressure, plus piston speed in feet per minute.

The amount of lubrication required for any engine is influenced by the quality of the lubricant, speed of engine, amount of work in proportion to size of engine, tightness of journals, correctness of alignments, finish of journals, truth of valve fane, perfection of cylinder bore and fit of piston.

Rule for finding contents in cubic feet of a cylinder of any given diameter: Multiply the square of diameter in inches by 0·7854, and this product by length of stroke in inches. Divide last product by 1,728 and result is contents of cylinder in cubic feet.

To determine the power of an engine multiply the area of the piston in square inches by the average pressure of steam in cylinder, and this product by piston's speed is feet per minute.

CHAPTER XIV.

DUST COLLECTORS AND DUST COLLECTING.

A few years ago the question of collecting dust or stive was of the utmost importance, and the subject received a great amount of attention from an engineering point of view, but recently it has lost a large amount of its significance. By recently is meant the period relating to the wide popularity of the dustless and air purifiers. Before their introduction the collection and deposition of dust particles was a matter of the highest consideration, and the market was flooded with all kinds of appliances to stop the waste which undoubtedly existed. It is not within the province of a text book to go back into what is already considered ancient milling history any more than just to mention one or two modes of dust collecting previous to the period alluded to. The writer suspects that not many modern mills are equipped with a stive room. This consisted of a large enclosed space or room, into which the various air trunks led from the fans. The principle was then, as now, viz., that a confined rush of air suddenly entering a larger compartment allowed for expansion, and by this means the particles which were perforce carried along the trunking were allowed to drop directly the impulse or suction showed signs of slackening, which, of course, occurred directly the larger area was reached, and the deposit was periodically cleaned out by hand labour. The march of science has, however, dispensed with these old time methods, and now the same ends are compassed by more subtle means, and that with a great saving in space and very little prime cost. The places where dust collectors are employed now may be summed up as follows: all cleaning machinery such as wheat aspirators, separators and graders, scourers and brushes, silos and elevators, and immediately before the wheat enters the break rolls. For all these purposes machines of a non-textile character are best, for reasons given in an early chapter, and which need not be further enlarged upon. Break and smooth rolls have an exhaust attached, and this current must needs end in a contrivance calculated to retain the light, filmy tissues which are drawn away with the warm air generated in those machines. Therefore, a collector of some kind is usually connected to make the necessary separation. Scalpers also utilise this means of getting rid of light floating substance, and some engineers go further and recommend a connection being made to all flour-dressing machinery to serve the same end; but as a rule it will be found sufficient if the other appliances named have an adequate amount of fan power attached without carrying the idea to extremes.

Formerly dust collectors were considered a source of danger, as were also stive rooms, but since the introduction of so many self-collecting devices the dust is deposited in small quantities as it is extracted; therefore, while the theory of dust explosion still holds good, there is now scarcely anywhere enough of it in any one place in the mill to cause alarm as when the whole of it was blown in a dense mass to a common resting place. The law governing the explosiveness of wheat or flour dust seems to be that of density—there must be a certain amount of it within a circumscribed area mixed with the oxygen and hydrogen of the atmosphere to cause ignition, but by the adoption of appliances splitting up this amount there is now never enough ordinarily to cause alarm. The writer goes even further than this, and ventures to assert that the collecting of this dust, whether from wheat or from broken particles, is now done in such small quantities by being spread over so many operations, that the danger which was impressed upon us ten years ago no longer exists at all.

CHAPTER XV.

MISCELLANEOUS TECHNICS IN BRIEF.

English wheat contains about 9,000 grains to a lb.

Australian wheat contains about 7,200 grains to a lb.

Russian wheat contains about 20,000 grains to a lb.

The first covering of the wheat berry is called its fruit coat and the inner layer the seed coat.

There are about 800 different varieties of wheat in the world.

It is estimated that every 100 ft. of 3 in. shafting absorbs 1 h.-p. in frictional resistance, as does also an ordinary sized roller mill.

An average yield from a wheat mixture will work out at

> 70 per cent. flour.
> $13\frac{1}{2}$ per cent. pollard or thirds.
> $15\frac{1}{2}$ per cent. bran.
> ___
> Total......99 per cent.

allowing 1 per cent. for dust, loss and evaporation.

In a business sense the quality of wheat is judged by its colour, outer form, smoothness, smell, the manner in which it breaks, by its feel, weight, and by measure.

The straw of wheat has four tubes, four knots, and each knot is situated at the point where two tubes join together.

An imperial bushel contains 2,218 cubic inches, and the

cubic contents of a silo is equal to its section multiplied with one-third of its vertical height.

The weight of a cubic foot of wheat varies with its quality and condition, but an average is reckoned at 46·82 lbs.

The contents of a bin or silo in cubic measure being known, divide it by 10¼ cubic feet and the number of quarters will be the answer; multiplied by 8 the figures will give bushels. Example :—

A silo of 120 feet by 40 feet by 84 feet $= 120 \times 40 \times 84 =$ 403,200 ÷ 10¼ $= 39,336\frac{24}{41}$ quarters $\times 8 = 314,692\frac{1}{2}$ bushels.

To obtain the weight in the same silo or bin, $120 \times 40 \times 84 = 403,200 \div 46·82 = 18,876,824$ lbs. ÷ 2,240 = 8,427 tons.

To find the height of a silo having a clear ground space of 100 feet by 30 feet to hold 8,000 quarters, proceed as follows :—$8,000 \times 10¼ = 30,750$ cubic feet ÷ 3,000 = 10¼ feet high.

To find out the capacity of bins or silos, proceed as under :—A silo is $30 \times 14¼ \times 12¾$ ft., how many quarters of wheat (500 lbs.) will it hold ?

ANSWER—Cubic content = 30 × 14¼ × 12¾

30 ft. = 360 ins.
14¼ ,, = 171 ,,
12¾ ,, = 153 ,,

360 × 171 × 153 = 9,418,680

1,728 cu. in. = cu. ft.   1,728) 9,418,680 (5,450 cu. ft.
8,640 · · ·

5,450
4
———
21,000 qr. cu. ft.

·7,786
6,912
———
·8,748
8,640
———
·1,080

1 qr. wheat occupies 10¼ cu. ft.

10¼ × 4 = 41 qr. cu. ft.

```
41 ) 21,800 ( 531 qrs.
     205
     ─────
      · 130
       123
     ─────
      · ·70
        41
     ─────
        29   ANSWER = 531 qrs.
```

RULES FOR ASCERTAINING DIAMETERS OF PULLEYS AND SPEEDS OF SHAFTS.

PROBLEM I. The speed of the driver, and the diameter and speed of the driven being given, to find the diameter of the driver.

Rule. — Multiply the diameter of the driven by its number of revolutions, and divide the product by the number of revolutions of the driver, the quotient will be the diameter of the driver.

PROBLEM II. The speed of the driven, and the diameter and speed of the driver being given, to find the diameter of the driven.

Rule.—Multiply the diameter of the driver by its number of revolutions, and divide the product by the number of revolutions of the driven, the quotient will be the diameter of the driven.

PROBLEM III. The diameters of the driver and driven and the revolutions of the driver being given, to find the revolutions of the driven.

Rule.—Multiply the diameter of the driver by its revolutions, and divide the product by the diameter of the driven, the quotient will be the revolutions of the driven.

Questions of this kind are frequently set at the examinations in flour milling, and they are not difficult when the formulæ given below is understood :—

Let $a=$ the speed of a shaft.

$b=$ the diameter of pulley on shaft.

$c=$ the speed of machine driven from shaft.

$d=$ the diameter of pulley on machine shaft.

Then $ab=cd$

And by transposing one of the terms of this equation, the value of each can be found. That is to say, if any three are given the fourth can be readily found, for

$$a=\frac{cd}{b} \qquad b=\frac{cd}{a}$$

$$c=\frac{ab}{d} \qquad d=\frac{ab}{c}$$

To make it more simple, the speed of a shaft multiplied by the diameter of the pulley upon it is equal to the speed of the machine driven from the shaft multiplied by the diameter of pulley on machine shaft.

$$\text{Let } a=240$$
$$b=20$$
$$c=160$$
$$d=\frac{240\times20}{160}=30$$

Again, Let $a=240$
$$c=160$$
$$d=30$$
$$\text{Then } a=\frac{160\times30}{240}=20$$

An example is now given without reference to Algebraic definitions, viz. :—A wheat brush is driven at 450 revolutions per minute; the diameter of the pulley on machine shaft is 10 ins.; the diameter of the pulley on the driving shaft is 24 ins. What is the speed of the driving shaft?

$$\text{Answer.}\frac{450 \times 10}{24} = 4,500.$$

$$24\ )\ 4,500\ (\ 187\tfrac{1}{2}\ \text{No. of revolutions.}$$
$$\underline{24}$$
$$210$$
$$\underline{192}$$
$$180$$
$$\underline{168}$$
$$12$$
$$12$$

Yet another way will be instructive, and that is, supposing a 30-in. pulley is making 180 revolutions, at what rate is the belt travelling?

ANSWER.—The speed of the belt is the same for practical purposes as the periphery speed of a driving pulley: therefore

$$30$$
$$3\cdot1416$$
$$\overline{}$$
$$94\cdot2480$$
$$180$$
$$\overline{}$$
$$7539\cdot8400$$
$$9424\cdot80$$
$$\overline{}$$
$$12\ )\ 16964\cdot6400\ \text{ins.}$$
$$\overline{}$$
$$1413\cdot7200\ \text{ft.}$$
$$\overline{}$$
$$= 1413\cdot72\ \text{ft. per minute.}$$

The heat of saturated steam at per lb. per square inch pressure above the atmosphere is—5 lbs. = 228 Fahr.;

10 lbs. = 240·1 ; 15 lbs. = 250·4 ; 20 lbs. = 259·3 ; 25 lbs. = 267·3 ; 30 lbs. = 274·4.

To find the safe pressure of steam boilers the rule is :— Multiply twice the least thickness of plates by 5,000, and divide the product by the diameter of shell in inches.

The slight swell in the faces of pulleys is usually about $\frac{1}{24}$ the breadth.

Pulleys should always be run with the curve of the arm facing the direction of the pull.

An ordinary sack of mixed wheat contains about three million of grains.

Sixty per cent. of the germ is said to be pure oil.

To a certain extent all cereals absorb moisture in winter and exude it in summer.

According to Professor Kick, no wheat contains less than 10 per cent. of water.

To take up half its own weight in water is considered a very good test for flour.

It is said that a grain of wheat has been known to yield —with its increase—16 bushels in three years, and in six years 200 sacks.

Higher belt speed, or larger pulley diameter, or both, increases the power of the belt for transmission purposes.

An ordinary purifier has seldom more than 4 lbs. of stock on the sieve at one time. It will operate upon 6 to 8 cwt. per hour.

If it were possible, all flour mill stock should travel by gravity from one machine to another.

The higher the speed of rolls the longer should be the bearings.

Every kind of wheat acts in some definite manner upon the resulting flour and bread, either by itself or by communicating its leading features to another wheat having affinity with it.

The number of corrugations on rolls ranges between 200 and 1,000.

The first break rolls are estimated to be when at work $\frac{1}{18}$ of an inch apart, and the last $\frac{1}{100}$ part of an inch apart.

The carbonic acid necessary to growing wheat is absorbed through the leaves.

The straw from an acre of wheat will draw from the soil 18 lbs. potash, 2 lbs. soda, 9 lbs. lime, 14 lbs. magnesia, 8 lbs. phosphoric acid, and 90 lbs. silica.

The weight of straw equals about 5 lbs. per bushel of wheat grown.

It takes about six weeks for wheat to mature after fertilisation.

Each grain of wheat takes up about $1\frac{1}{4}$ square in. roll surface on a run of five breaks.

There are varieties of wheat with 19, 21 and 24 spikelets, each spikelet contains five flowers, which yield two, three, and sometimes four grains each.

The capacity of a break roll is simply the weight of a single layer of wheat equal to the roll's surface area multiplied by its revolutions per minute or hour.

To calculate horse-power in water for turbines:—Strike an average depth of water, multiply this by the breadth of water, multiply again by its velocity per minute, multiply this total by the weight of 1 cubic ft. of water, multiply again by the fall available, divide by 1 horse-power, viz., 33,000 lbs., and the result will guide towards a decision.

A comfortable working strain for single belts is 50 lbs. for every inch of breadth, and 80 lbs. for a double belt.

FLOUR AND WHEAT VALUES.

Value (say) of a quarter of wheat 35/-
  „   „   „   cental   „   7/-

Value of 400 lbs.   ...   28/-=70 per cent. flour,

and this is a rough guide to the price at which a straight run grade will sell relatively.

RULE FOR FINDING THE LENGTH OF A ROLL OF BELTING.

Take the over-all diameter and add to it the diameter of the hole in the centre of the roll ; then divide the sum by 2 to find the mean diameter ; this multiplied by 3·1416 (or 3$\frac{1}{7}$) will give the circumference. Then multiply this by the number of " laps," and the result is obtained in inches, and by dividing by 12 the length of the roll is obtained in feet.

Every inch in width of good double leather belting, travelling at 500 ft. per minute, will transmit 1 horse-power.

Every inch in width of good single leather belting, travelling at 800 ft. per minute, will transmit one horse power.

To calculate the capacity of a worm, take the diameter, square it, multiply by ·7856, divide by 2, multiply by the pitch and revolutions per minute, and then divide by 2218 ; the answer will be the number of bushels conveyed per minute. Multiplied by 60 it will give the quantity per hour.

The maximum speed of worms is about 100 revolutions per minute, but if possible they should not much exceed 80 revolutions, as this is their most effective speed.

The proper pitch of a worm should be about $\frac{2}{3}$ of its diameter.

A 5-in. worm with a 2-in. pitch will carry about 40 bushels an hour running at 80 revolutions per minute. Example given according to above formula :—

$$\begin{array}{r} 5 \\ 5 \\ \hline 25 \end{array}$$

·7856

19·64 area of circle

÷ by 2 = ½ area 2/19·64

9·82

160 pitch and revolutions

58920
982

2218) 1571·2 (·708 bus. per minute conveyed
15526

18600   ·708
17704   60

· ·856   42·48

Answer—42·48 bus. per hour.

It will also be well to just give an instance of the capacity of belt conveyors, and, taking one 12 ins. wide as being fairly representative, it may be said to carry an average of 2 lbs. weight of wheat per foot, travelling at a speed of 500 ft. per minute, and this equals 60,000 lbs. per hour, or, worked out in figures the answer will be as under :—

2240) 60,000 (26·78 tons per hour.
4480

15200
13440

·17600
15680

·19200
17920

·1280   Answer, 26·78 tons.

RULE TO FIND THE HORSE-POWER THAT ANY GIVEN WIDTH OF DOUBLE BELT IS CAPABLE OF DRIVING.

Multiply the number of square inches covered by the belt on the driven pulley by one-half the speed in feet per minute through which the belt moves, and divide the product by 33,000. Quotient will be the horse-power.

RULE TO FIND THE WIDTH OF BELT.

Multiply 33,000 by the horse-power required, and divide the product—first, by the length in inches covered by the belt on the driven pulley, and again by half the speed of the belt in feet per minute. Example:—

> Speed of Engine, 50 revolutions.
> Driving Pulley, 24-ft.
> Driven Pulley, 9-ft.
> Belt, 30-in. wide.

Diameter of Driven Pulley, 9-ft.=28-ft. 3-in. circumference.

Half, 14·15, or
169·8 × 30-in.=5094. Half speed of belt, 24 feet, or
Circumference, 75-ft. 4 in.

50
___
3770

33,800   Half, 1885 × 5094=
___

=9,602,190=291 horse-power.
Engine for this belt, &c., as above.

291 horse-power requires size of belt.
169·8
291 × 33,000——— ———
9,603,000=56,555=30-in. width of belt.

Lubricating has for its object the reduction of friction between moving surfaces.

For heavy machinery bearings a heavy lubricant of high viscosity should be employed, while thinner oil may be employed on lighter and more delicate movements.

Principal shipping ports of foreign wheat exported to England :—

Wheat.	Ports.
Australian ...	...From Melbourne and Sydney.
Californian	,, San Francisco.
Indian	,, Bombay, Calcutta, Kurrachee.
South Russian	,, Azov, Odessa, Taganrog.
Argentine	,, La Plata, Rosario.
Minnesota, Duluth, &c.	...To the St. Lawrence, or by the Canadian Pacific Railway to the ports on the Atlantic.
Manitoban ...	...Through the Lakes to the St. Lawrence, or by rail to Atlantic ports.

Ordinary flour and corn sacks weigh 1 lb. for every bushel of capacity.

For belts which are to be crossed the extra allowance for length when calculating is about $1\frac{1}{2}$ in. for every 10 ft.

It takes 400 lbs. of clean wheat to make 70 per cent. of flour.

   ,,   395   ,,   ,,   71   ,,

   ,,   388   ,,   ,,   72   ,,

A suitable distance apart for elevator cups is 12 to 15 ins.

The speed at which wheat elevators will discharge easily is 300 ft. per minute.

To find the speed of cog-wheels, substitute the number of cogs for inches diameter and proceed as with pulleys.

To take percentage of finished flour the simplest method is to do so by a rule of proportion, viz., add up the amount of wheat ground, reduce to pounds weight, do the same with the flour and offals, add two ciphers to the latter, and divide by the former for each separately; that is, flour first and offals afterwards.

Example—1000 lbs. weight in wheat ground

out of which is made   700 lbs. Flour
                      138 lbs. Pollard or Thirds
                      152 lbs. Bran

Total   990

Wheat   1000)700·00(70 per cent. Flour
            7000

           0000

Wheat   1000)138·00(13·8 per cent. Pollard or Thirds
            1000

           3800
           3000

           8000
           8000

           0000

Wheat   1000) 15200 (15·2 per cent. Bran
            1000

           5200
           5000

           2000
           2000

           0000

Making a total of—
        Flour......... 70·0
        Thirds ...... 13·0
        Bran .... 15·2

                99·0
        Loss .... 1·0

              100·0

It is estimated that break rolls are at a pressure of about 600 lbs., and some smooth rolls as high as 1,000 lbs. per roll spindle.

Openings per square inch in silk—

0000 =	324	meshes	(about)
000 =	529	,,	,,
00 =	841	,,	,,
0 =	1444	,,	,,
1 =	2401	,,	,,
2 =	2916	,,	,,
3 =	3481	,,	,,
4 =	3969	,,	,,
5 =	4489	,,	,,
6 =	5626	,,	,,
7 =	6724	,,	,,
8 =	7396	,,	,,
9 =	9604	,,	,,
10 =	12100	,,	,,

Purification is only absolutely perfect (or can be made so in practice) when every particle fed on to the machine is of exactly the same size.

Microscopic examination shows that even the finest and whitest flour contains bran powder.

A quarter twist belt loses 10 to 25 per cent. of its duration of contact on account of the inclined deflection of the fold as it leaves the pulley; and the correct alignment or relative position is that the centre of the driving pulley should be set exactly in line with the outside edge of the driven pulley.

Many students are somewhat hazy on cubic measure, and a little light on the subject may be of assistance. A cube is a solid body, and contains length, breadth, and thickness, having six equal sides.

A cube number is produced by multiplying any number

into itself twice, instance, 64 is a cube number of 4 : thus $4 \times 4 \times 4 = 64$.

> 1728 cubic ins. = 1 cubic ft.
> 27　,,　ft.　= 1　,,　yard.

SUPERFICIAL MEASURE.

> 144 square inches = 1 square foot.
> 9　,,　feet　= 1　,,　yard.

A bushel of Linseed, on an average, weighs 53 lbs.

,,	Malt	,,	,,	,,	38 ,,
,,	Beans	,,	,,	,,	64 ,,
,,	Peas	,,	,,	,,	64 ,,
,,	Oats	,,	,,	,,	40 ,,
.,	Barley	,,	,,	,.	50 ,,
,,	Rye	,,	,,	,,	53 ,,
,,	Wheat	,,	,,	,,	60 ,,
,,	Flour	,,	,,	,,	56 ,.

The diameter of a shaft cubed = its horse-power.

The proper place to stand when putting a belt on is on the opposite side, and pull the belt on the pulley as it travels towards you.

> A peck loaf weighs ...　...　$17\frac{1}{2}$ lbs.
> A quartern loaf ,,　...　...　4　,,
> A peck of flour ,,　...　..　14　,.

CHAPTER XVI.

EXAMINATION QUESTIONS.

The following selection from questions set at recent milling examinations will be found of some interest to all millers. Independent answers are subjoined.

You are required to make a flour which would match a straight run made from sound No. 2 Red Winter wheat. Give alternative wheats, or mixtures of wheat, in the order of your preference, and, among other points, consider the colour of the crust of the bread produced therefrom, and the satisfactory working of the doughs.

Answer.—The following wheats will be something like No. 2 Red Winter : Canadian, dry English, Rostock and Dantzic. These will be equal to the Winter on a dry-cleaning process. On a washing plant you will be able to use such wheats as Danubian and Indians in a mixture.

No. 2 Red Winter   ... 1. Canadian Red.

"   ... 2. { 1 Dry English.
                      { 1 Mild Baltic.

"   ... 3. { 1 River Plate or Kansas.
                      { 1 English or Baltic.

"   ... 4. { 1 Canadian or English.
                      { 1 Danubian or Mild Indian.

Is the percentage of gluten a flour contains an index of any value as to its capacity for making big bread ? Give several examples illustrating your answer.

Answer.—No; for the following reasons:—Gluten being a composite body, composed of two constituents—viz., gliaden and glutenin—it depends upon the proportions in which these two bodies are present in the gluten more than in the quantity. For example, take the following wheats:—

Name of wheat.	% of gluten dry.	Strength.
Duluth	12 to 16 ..	High.
Hungarian .. .	9 „ 18 ..	„
Australian .. .. ·	8 „ 10 ..	Good.
New Zealand ..	9 „ 10½ ..	Low.
Persian	8 „ 10 ..	„
Californian	10 „ 11 ..	„
Indian	11 „ 12 ..	Medium.
English	12 „ 14½ ..	Fair.

Percentage of wet gluten, 2½ to 8 times the amount of dry gluten.

The mixture you have to use contains Walla Walla, and a proportion of a sort such as Kubanka or Canadian goose, or hard Taganrog. Should these two wheats be conditioned together? Would such treatment affect the quality of the flour produced? Give the reasons for your answer.

Answer.—No; for the following reasons:—The Walla Walla, although a dry, brittle wheat, will not stand the water well; it absorbs it freely, while the other wheats named, especially the Taganrog, require a considerable amount of water, and time in the water, to bring them into anything like milling condition.

The method of conditioning those wheats would not affect the flour to any great extent. The Walla Walla would perhaps suffer a little, but as it would not be used for any strength that it had, the excess of water that it would get would do very little harm to the strength of the resultant flour. The strength is the chief thing to guard in the conditioning of several wheats together, to see that none of them get too much moisture, for in the presence of excess

K 2

of moisture the gluten of flour is rendered in part soluble,
and it also loses in elasticity; while, at the same time, the
starch is degraded more or less into maltose and dextrin
by diastases.

How would you ascertain the percentage of moisture
added to or extracted from your various wheats in passing
through a washing and conditioning plant?   Give simple
sketch of apparatus used, if any.

Answer.—By means of an automatic weigher, weighing
the wheat before it entered the washer, and after it had
passed through the conditioner.   By this method you
would be able to ascertain more correctly the exact gain or
loss which has taken place during the process.   You can
also find out the percentage of water present in wheat by
means of a hot water oven, but I think this more the work
for a chemist than a miller, therefore I should leave it
alone.   There is also the method of weighing the wheat so
many pounds to the bushel as it passes through the conditioner, but it does not give the exact gain or loss,
although it is a bit of a guide.

Describe the methods and principles employed in a well-
known "dustless" sieve purifier for separating air-borne
particles from the wind currents which have extracted them
from the stock to be purified.   Has the quantity of feed on
the sieve anything to do with the efficiency of the "dust-
collecting" apparatus?

Answer.— The methods are various, but the principle is
the same in all modern purifiers, that is, air contraction
and expansion and gravitation.   Contraction applied to air
currents means forcing the same volume of air through a
smaller space than that through which it had passed in the

first instance. It manages to do this by travelling at a greater velocity. Expansion is the reverse. More room or space being allowed, the air current naturally spreads itself out, and does not travel so fast ; therefore, the air has less power over the dust particles, which gradually drop or settle down into the collecting troughs, platforms, &c., while the air itself is discharged from the fan practically free from dust.

The quantity of the feed on the purifier sieve will affect the dust-collecting apparatus in this way. If the machine is over-fed on certain stock you will be forced to carry more wind on the machine if you want to get the throughs of the sieve clear, and the more wind you have on the machine the more difficult it is to separate the dust particles from the air.

Are you in favour of using reduction rolls over 40 ins. in length ? Give reasons for your preference or objection.

Answer.—No. At the present time fine dressing is the order of the day, and the tendency is still upwards ; therefore, the grinding will have to be equally as fine, and to get this you will have to set your rollers closer, with the result that if not very carefully handled the rolls will be liable to get off at the centre, thus allowing the stock to get through without being sufficiently reduced. The greater the length of the rolls the worse they are in this respect. Another reason is that except in very large mills it would be difficult to get sufficient feed of the same quality to feed rolls over 40 ins. on the lower reductions, and mixed feeds rarely give satisfaction, so that where smaller rolls are used stock of all the same quality can be ground by itself to the best advantage. An advantage that large rolls have is this :— A 50-in. roll only requires the same attention as a 25-in. roll, thus if you use two 25-in. instead of one 50-in. you will have double the work, and it is easier (presumably) to keep two ends alike than it is four. Prime cost is also a feature.

Apart from sprinklers give your ideas as to the best methods of protecting mills against fire whilst working, and also whilst standing at the week end.

Answer.—To protect a mill against fire I should say keep it clean and in good repair, no bearings out of order, or bad running belts, all fan bearings clear of the woodwork, a systematic examination of all heavy and fast running bearings. No dirty waste or damp sacks to be allowed in the mill, no naked light, electric wires to be kept away from damp places. The screen room wants watching particularly. Appliances in case of fire. Plenty of fire buckets always kept full, fire queen, hydrants and a length of hose on each floor, and the men trained how to act in case of fire. At the week end a good watchman, one that will give extra attention during the first six hours after the mill has been stopped; there should also be communication with fire or police station.

A fresh sort of wheat of unknown or uncertain quality has to be introduced into your mixture in any proportion found most desirable. Say, in detail, how you would test it for quality, and how, having regard to that, you would arrive at the proportion you would use.

Answer.—The only safe way to ascertain the quality of flour from an unknown wheat would be to put a trial lot on to the mill by itself and have the flour baked. The result will be a guide, more or less, what change you can make in your wheat mixture. The proportion you would use would have to be governed by the price paid for the wheat and the price flour had to be sold at. There are several other ways of testing flour, but baking it against, say, the kind it supplants is, in my opinion, the only sure method. Others, such as gluten tests and colour tests, are misleading.

Forty per cent. of the mixture you are milling consists of a good relatively clean wheat of high natural weight. For that it is proposed to substitute another lot of wheat containing flour of the same quality, but of light natural weight, with, say, 6 per cent. more screenings and 2 per cent. more dirt than the original lot. Say how the substitution of the lighter wheat for the better would affect the working of the mill and its working expenses. How much would the extraction of the excess of screenings and dirt add to the cost of the wheat? Set out the calculations by which you arrive at your answer.

Answer.—By the substitution of the lighter wheat the mill would fail to give so good a yield (percentage), and it would also curtail the capacity of the mill a little. It would also not stand as much water, and as far as it altered in these respects it would be the difference in the expenses.

Assuming that the wheat mixture in the first instance cost 30s. per qr., 500 lbs. (actual figures not given), and the screenings 4· per cent., flour 71· per cent. from clean wheat, average price 25s. per sack, offals average 100s. per ton. All screenings ground up and mixed into offal, conditioning to make up for any loss in weight by evaporation, &c.

Mixture No. 1.

	lbs.				
Dirty wheat 1 qr.	500 cost 30s.				
4· per cent. screenings	20	at offal price	=		price including offals.
71· per cent. flour (Clean wheat)	340½	at 25s. per sack	=	30	5⅞·¹⁴/₁₄
29· per cent. offal (Clean wheat)	139½	at 100s. per ton	=	7	1¼·¹⁶/₁₁₅
100	500		Total	37	6¼·⁷

Mixture No. 2.

12· per cent. screenings	60 at offal price	=	price including offals.
71· per cent. 29· per cent. (Clean wheat) flour / offal	312⅖ at 25s. per sack 127⅗ at 100s. per ton	= =	27 10½·9 8 4½
100	500 Total		36 3¾·9

Difference in cost per qr. of wheat   1   3¾·9

P.S.—If screenings were sold as such the difference would be rather more.

Is the hardness or mellowness of wheat any index of strength or weakness? Give examples illustrating your answer.

In this question we will treat of both conditioned and unconditioned wheat. Taking wheat first in its natural state, we find that Indians, Egyptian, Chilian are amongst the hardest wheats, but are by no means of any account for strength, as we find that good English upon this point is quite equal to unwashed Calcutta, but to wash and condition these two wheats the Calcutta would show superior results.

The same thing holds good in the case of mellow wheats, as Oregon, New Zealand, Red Winter, Dantzic, and Konigsberg, all fairly mellow wheats, yet not to be compared for strength with Spring American, which is no harder than some of them. Of course, in the case of washed and conditioned wheats, we should find a considerable difference in some of the above, as in the case of some of the very hard wheats after washing we find the gluten and general strength very much improved. Finally, I should say that the hardness of grain is no criterion of its strength, but that on the other side of the question no soft wheat has any claim to strength, but we also find that all soft or mellow wheat is vastly superior from a colour point of view in the resulting flour.

Can damp, smutty home-grown wheat be washed and made fit to grind without making it weaker ? If you think it can, say how you would do it.

In reply to this question I should say "No," as it requires a good deal of water to properly clean it, but it can be so dried that the strength will be so little affected that it can be made of good value for milling purposes. If, however, this wheat is very damp, and contains about 16 per cent. of moisture, it can perhaps be properly washed and dried so that it will only contain about 13 or 14 per cent. of moisture when conditioned, but this can only be done with a good plant having ample whizzing and drying capacity.

The way I should do it would be as follows :—The wheat, upon arrival at the mills, should be at once put over a warehouse separator with a good exhaust to take out as much of the unbroken smut as possible so that it would not get more broken in its transit to the silos and cleaning plant. Upon starting to clean this wheat I should make it my endeavour to clean it as far as possible from all smut balls and adhering smut before washing, as once let smut balls get wet, they cannot be taken out at any time in the later processes of milling. The wheat having arrived at the milling separator should be met with a powerful exhaust, both upon entering and leaving the machine, rather lifting some small wheat than leaving in any smut balls. From milling separator I would send it to cockle and barley cylinders, then to scourer, where it wants again to be carefully exhausted and scoured as severely as possible without breaking the wheat. From here to the washing plant, where you must give it plenty of water and a good washing and rinsing, then whizz it for all your machines are worth. Then to hot air dryer with plenty of hot air about 140 to 150 degs. Fahr., and up in the cold section of the dryer. Give it all the cold air you can get on to it, and have it leave the dryer properly cold, and not over $1\frac{1}{2}$ lbs. per bushel lighter than its natural weight when it left the scourer. From dryer to condition bins, let lie for eight hours, then brush and send to mill bins for grinding. It will be found equal to your average English mixture.

Describe in detail a " Scratch " roller mill. Say whether you would use any, and, if so, how many, and in what position or positions, in an 8-sack plant?

The details of any makes of scratch roll from that of break or smooth rolls from the same makes is only upon the differential speed and the corrugation of the rolls. The speed for a 9-in. roll would be as follows:—Fast 300 revolutions; slow roll, 100 revolutions, or as three to one.

The corrugations are from 28 to 32, according to the class of wheat worked.

Great advantages for any mill are to be derived from a scratch system, as it enables the miller to use a better system of breaking, and also to get more and purer stock at the head of the mill, with less impure stock upon the tail, and brighter flour all round.

Upon an 8-sack plant I should use two systems of scratch rolls, one pair of 24-in. upon each. The first or X system would treat the tail sheets from coarse and fine semolina purifiers. From the roll this would go to a centrifugal clothed with No. 70 F wire, overtails to purifiers, throughs to BMR. The second or Y scratch system would work upon the overtails of the coarse and fine semolina purifiers, over-tail and cut off upon X purifier to Y scratch roll, then to IV. B^k scalper, through No. 36 fine wire to IV. R^e dresser clothed with 11, 12, 8, overtails to last purifier. With plenty of break surface and a good scratch system the break flour can be brought down to about 14 per cent., which means more patent flour, more money, and better results all over the mill.

From which of the following can you obtain the broadest straight-run bran :—

(A) Well-conditioned No. 1 Northern Manitoba wheat.

(B) A mixture of 60 per cent. No. 1 Northern Manitoba, 20 per cent. Australian, and 20 per cent. imperfectly conditioned English ?

Give reasons for your answer.

(A) I should expect to get the best all-round bran from the first-named, as the wheat being in first-class condition and all of one toughness would not cut up so badly in breaking as with the later mixture.

(B) This mixture is of an unequal hardness, and in the endeavour to clean up the badly conditioned English upon the break rolls the other wheats would be quite clean before reaching the last break and would get badly cut up upon this break and give an unequal bran.

What are the effects of allowing break rolls to get dull upon—

(A) The " patent " flours ;

(B) The flours made by the breaks, excluding the last ;

(c) The flour made by the last break ;

(D) The bran ;

(E) The capacity of the mill ?

(A) The effects upon the patent would be that it would be poorer quality, and there would be less of it than with sharp rolls, as the semolina and middlings would not be so clean cut or dusted, and would contain more impurities in the form of cut up bran, and cause worse purification.

(B) The flour from the breaks would be of better colour, and there would be more of it, would possibly reach 20 per cent.

(c) The flour from the last break, if there was any endeavour made to clean the bran, would be very brown and full of bran powder.

(d) The bran would be poor, irregular and badly cut up, and very ragged and torn.

(e) The capacity of the mill would be greatly reduced, as the breaks, to begin with, would not be fit for their full feed. The stock to the purifiers would be badly dusted, giving poor results, with heavy tins and overtails and low percentages all round.

The feed of your first break is discharged from a grinding bin by means of an adjustable slide. Without any alteration of its adjustment, you find the quantity of feed on the plant varies considerably. From what cause, or causes, would this arise, and to obviate such variation would you recommend the adoption of any apparatus to feed the plant with any given number of pounds of wheat per minute quite regularly?

This may be caused in various ways, but it may be set down to one of the three following:

First, some block in the shoot, as pieces of straw, &c.

Second, the condition of the wheat, as it is well known that wheat that is put on to the mill in three or four different conditions in the 24 hours will each time make an alteration in the feed of the plant. Wheat that is moist does not run so freely as wheat that is dry.

Third, the state of the atmosphere. If the air is full of moisture wheat does not run so freely as when it is dry and clean.

I would not go in for putting in a weigher to feed so many pounds to the mill, as I have always found that when wheat is damp or in a damp atmosphere the mill cannot take the same feed. Instead of a slide feed I should prefer a mixer driven from the mill to set the feed by.

What effect, or effects, has four months' warehousing upon flour milled in the United Kingdom? Has the grade of the flour, or the time of year, to be considered in connection therewith?

Flour after being milled and warehoused for four months in this country has a great tendency to go wrong. Its only chance of keeping for this length of time is for it to be milled from all foreign wheat, and that wheat being unwashed. Dry milled flour warehoused for four months would tend to get mitey and have a bad smell, and would lose in weight. Against this this can be said, that it would be whiter in colour. If during the winter, this flour would not take any harm through keeping. Flour made from washed wheat and conditioned so that moisture was added to the berry would not keep in good condition for four months, as it would have a tendency to get into small lumps, which would soon get yellow in the centre, and have a bad smell. This is more pronounced in the break and low grade flours, especially if it was milled during warm weather, as under these conditions break flour would get quite hard in the centre of the bag, and lose all the strength it had by some kind of fermentation having been started.

Describe the various conveyors for moving wheat, flour, and offals; state which you consider most suitable for each.

Answer.—The usual methods of moving wheat, flour and offals are either by continuous or paddle worms or a band conveyor.

To move wheat any distance a band conveyor is far preferable to any other appliance, as it consumes but $\frac{1}{18}$ of the power to do the same work as a worm; moreover, a band conveyor gets through its work much faster, running, as it does, 500 ft. per minute.

For short distances a continuous worm is best except for wet wheat, in which case it is advisable to use a paddle worm.

To move flour from dressing machines to packer, use paddle worms, as they mix the flour as it travels.

To move offals use a continuous worm unless the distance be great, when a band conveyor running 400 ft. per minute may be recommended.

Describe the various methods of hoisting used in flour mills; which do you prefer, and why?

Answer.—Friction hoists are possibly the most to be recommended, because they

  1st. Require little space.

  2nd. Are not likely to get out of order.

  3rd. Are simple and easy to work.

  4th. Have a small first cost, and their wear and tear is small.

A hoist in very common use is driven by a belt; the belt hangs loose on the hoisting barrel and is put into motion either by lifting the barrel to tighten the belt, or else by a tension pulley; this hoist has much to recommend it, its chief fault is that the belt stretches and is liable to break. There are also conical and disc hoists on the market which do good work. In choosing one, always take care that the hoist is of simple design, is easily worked, and has few wearing parts.

What are the usual speeds for running the following :—
Roll main shaft, smutter, rolls, centrifugals, purifiers, reels,
elevators, worms, band conveyors, and sifters ?

Answer.—Roll main shaft, from 125 to 150 revolutions
per minute. Smutter, 450 to 600 revolutions per minute.
Rolls (fast roll), 600 to 800 ft. per minute ; ditto (slow roll),
230 to 300 ft. per minute. Centrifugals (beaters), 160 to
250 revolutions per minute, according to diameter. Purifiers, 450 to 500 revolutions per minute. Reels, 25 to 35
revolutions per minute. Elevators, 200 to 300 ft. per
minute. Worm conveyors, 60 to 80 revolutions per
minute. Band conveyors, 400 to 500 ft. per minute.
Sifters, 400 to 500 revolutions per minute.

How much power does it take to produce 10 sacks of
flour per hour in a well-equipped plant, and how much of
this is absorbed by the rolls?

Answer.—In a well-designed 10-sack plant it would take
from 70 to 80 h.-p. from the clean wheat to finished flour
and offals ; of this, fully two-thirds would be absorbed by
the rolls—i.e., from 47 to 54 h.-p. The wheat cleaning
would take from 25 to 35 h.-p.; but in giving these
figures a liberal allowance must be made, as possibly no
two mills are alike in this respect ; so much depends upon
the class of wheat ground, both for washing and conditioning,
as well as grinding and dressing. Possibly it is not far
wrong to say 10 h.-p. per sack will cover everything in
large, well-designed mills, and half of this power will be
wanted for the rolls. The reduction rolls take more power
than the break rolls in about the proportion of three to two.
But it must be borne in mind that in most small mills, and
in many others where the machines have been installed in
an unsuitable building, it will take 12 h.-p. per sack, and
even more.

Enumerate the daily duties of a rollerman, a purifierman, and a silksman.

The rollerman should oil round before the start, and see that all drives and belts are in order. As the stock arrives in the hoppers of the various breaks and rolls, he should set them to work and never allow a break or smooth roll to run empty.

After seeing the breaks are fed and working properly, he should do likewise with the smooth rolls, also feeling the ground products to satisfy himself that the rolls are set true and at correct distances apart.

He must see that the wheat coming to the breaks is well conditioned ; that the exhaust to each machine is sufficient (stive-boxes want periodically cleaning out). If the feed is badly dusted he must inform the silksman, and if the smooth roll feeds are unsatisfactory, the purifierman must be told. If he is not satisfied with the finish at the end of the mill he should lessen the feed, and see that any errors are adjusted to remedy this.

After a couple of hours' run he should start sampling, taking the feeds and ground products of each machine and laying same on boards, which should be fixed at the windows. From these samples he can judge if his work is at its best.

Lastly, he should oil round occasionally, sweep down machines and brush up floors.

The purifierman should first oil round, and place all drives and belts ready for a start, then stand by and see the drive starts properly. As the feed arrives on the machines he must carefully adjust the fans, and by examining the tails and last sheets find out if they are lifting too much or not enough.

After seeing that the silks are tight and all sieves are running properly, and not throwing to one side, he must examine the cut-offs, and adjust accordingly. He must take care the grading of the semolina and middlings is properly done, and that same is free from dust.

After running a couple of hours he should commence sampling, taking about 2 ozs. of the feed, each sheet—the cut-off—tail and aspirations, and laying them on the sample boards and smoothing with a spatula, he should be able to see if his machines are working to their full advantage, and any holes in silks will at once be detected.

Lastly, he should oil round occasionally, and sweep up.

Before starting, the silksman should look round the covers and patch up any holes, then oil round and fix belts and drives, see that the divides are properly set for the brand of flour required.

After running a few hours he should start taking a sample of flour from each machine ; these must be carefully examined, and if found branny the hole should be traced and mended; if time allows he should wet up small portions of each flour ; samples of the dunst and tails of each centrifugal should also be taken.

The silksman must look after the percentage of each flour taken off. Any variation in grades he should report to the rollerman.

What are the causes which contribute to making more than the usual or proper percentage of break flour, and how can they be avoided or obviated ?

One of the most common causes of break flour being made in excess is dull break rolls, which, instead of cutting as they would if sharp, simply bruise the product passed through them ; hence the break flour instead of middlings.

The remedy is to have them re-corrugated.

Badly and irregular conditioned wheat also tends in the same direction, the moisture possibly having penetrated to the cells of endosperm, thus weakening the affinity of the particles with each other, and causing them, when under treatment on breaks, to separate into floury particles. Remedy: Overhaul conditioning plant, also the men that they know what is required, and how to do it. Bad handling of break rolls may also increase break flour; for instance, too much work may be done on the earlier breaks. This makes the rollers then approximate somewhat to the stone system, with what result we all know.

Then, again, too little may be done, leaving thick and good material to be acted upon by the later breaks, where it is only possible to make small middlings and break flour.

Slipping belts may possibly account for the evil, as instead of maintaining the proper differential speed between the respective rolls, the rolls tend to run together at the speed of the roll having the tightest belt. This gives a pure bruise to feed. The remedy for above is to go round rolls one by one from first to last, then back and repeat, noting feed, condition, treatment received, belts, &c., adjusting and tightening where required.

Excessive feed will also tend to increase the quantity of break flour, so that the feeds of respective rolls, instead of being a single thickness, are more than this, thus causing the crushing action so effective for the making of break flour.

An increase of soft wheats, such as English, in mix, is almost sure to cause an increase of break flour unless the wheats are very carefully blended and conditioned.

The product from rolls (belts of which are slipping and thus bruising or rolling product) is simply break flour and bran afterwards, and very little else will result after being once rolled.

Mention some of the causes which tend to make the brands of flour variable on the same wheat mixture.

Principal causes: (1) Washing, (2) engine speed, (3) feed regularity, (4) weather.

(Minor causes, affecting grades and divisions only. Bursting silks, chokes, slipping of purifier belts, unclean silks, &c.)

(1) Washing: This department if not skilfully handled will cause a fluctuation in the hardness of the grist, insufficient drying or excessive washing resulting in "softness," insufficient washing or excessive drying resulting in "hardness." A hard grist is more granular than a soft grist, and the easy dressing out will make the flour "grey." A soft grist is affected *vice versa* with an excess of offal and low-grade flour.

Engine speed: This is important in that it should be uniform, because a reduction in speed will cause an accumulation of feed and inefficient dressing out on all sifting machines, thus sending more than due proportions to the tail of the mill. Excessive speed will act *vice versa* and will make the dressing too bare.

Feed quantity: All machines are made with a certain capacity, and, if they have not enough feed, the work will be overdone, greyness appearing in the flour. If they have too much feed, the work is underdone, increasing the quantity of the low-grade at the expense of Patents.

Weather: A dry crisp atmosphere with a high barometer will quicken the dressing out and tend to increase the quantity, but reduce the quality (colour) of the best grades. A cloudy, humid atmosphere, with a low barometer, will obstruct the dressing out, and tend to reduce the quantity of Patents, making an excess of low-grade.

You are required to clean and condition ready for grinding the following wheat stored separately in granaries or silos, and to deliver them to the grinding plant, mixed in the following proportions :—

25 per cent. No. 1 Northern Duluth.

25   ,.   Hard Winter.

10   ,,   Australian.

20   :,   F.a.q. R/SFe Plate.

20   ,,   English, F.a.q. and dry.

———

100

You have at your disposal a complete wheat-cleaning, washing and conditioning plant, and are not restricted as to the number of bins or as to time in which to do the work. State at what point, or points, you would make the blend, how you would clean and condition the wheat, and specify approximately the moisture of each wheat before cleaning and after conditioning.

Wash the Duluth, Hard Winter, and Plate together, and as the feed passed on to the dryer and conditioner meet it with the Australian and English. Pass the mixture into one of six large conditioning bins, large enough to hold a day's grinding. To fully explain this, we will suppose that we grind out of bin No. 6. An elevator connects each bin with another. We elevate from 4 to 5, from 3 to 4, and from 2 to 3. No. 1 should be filled with freshly washed and conditioned wheat on one day and No. 2 the next, emptying these into No. 3 in turn. Grind out of both No. 5 and No. 6, so that No. 4 would be emptied alternately into No. 5 and No. 6. This is the very best system if bin space and capital will allow of its being done, and the wheat is always in first-rate grinding condition. I would clean the wheat through the medium of separators, graders, cylinders and brushes—no scourers—and would wash, whizz, and send them to hot and cold air chests, as already mentioned. Regarding the moisture, I submit the following table :—

Before cleaning. Moisture.	After conditioning.
Duluth, 10 per cent.	14 per cent.
Australian, 9 per cent.	11 per cent.
Hard Winter, 10 per cent.	13 per cent.
Plate, 9½ per cent.	15 per cent.
English, 14 per cent.	12 per cent.

In a five-break mill, working with its wheat mixture of the usual condition and quality, you find the third and fourth breaks working hot, and the "throughs" soft and greasy. What would you think wrong, and what would you do to put it right?

Anything wrong in this department necessitates a look at all the preceding machines. If the third and fourth breaks were working hot and making soft and greasy middlings, they are very likely struggling to get rid of some accumulation, or are in themselves set up too close. Most probably the former is the case, and an examination of the second break roll will in nearly all cases discover the mischief, and that would be set right by making this second break roll do more work. In this and similar circumstances it means that the proper proportional amount of work allotted to each pair has been violated, and so the system gets disorganised, and in trying to treat material which the rolls or succeeding machines are not set up to do, they bring about an entirely different result.

For the milling of a mixture consisting of two-thirds foreign wheat, washed and conditioned, and one-third English :—

(A) What I.H.-P. would be required to operate the entire plant, exclusive of Silo-house Machinery?

(B) Of this, how much would the wheat-cleaning, washing and conditioning absorb?

(C) How would the substitution of sieve-dressing for centrifugals affect the amount required?

(D) What effect would the lengthening of reduction roll surface per sack of flour made have on the power consumed?

(A) About 10 h.-p. per sack per hour as a minimum.

(B) Somewhere about 4 h.-p. per sack per hour, calculated on the flour output.

(C) Would result in a saving of at least 10 per cent. of total power used.

(D) Very little. It must be understood that for every pair of rolls added to the reduction side the pressure is slightly eased on all, and this just about balances the extra machines added.

Give your ideas as to the distribution, etc., economy of labour, in a 15-sack plant. Say how many men you would need to work the plant.

Two rollermen,
Two purifier men,
Two silkmen,
Two screenmen,
Four packers,
Two engineers, and
Two spare men.   There would also be
One night foreman,
One day foreman,
One manager = 19 men in all.

Specify the essentials for the carrying out of good purification.

Middlings must be as large as is consistent with the system, and there must be as many as can be had. The purifier is built upon a certain plan to deal in a certain way with certain prepared material. Let the material be not what the purifier is built to deal with then all purification —efficient purification—is out of the question. Dirty wheat, dusty middlings, ill-conditioned wheat, and stock of greatly divergent sizes are against good purification. Briefly the essentials are as follows:—

1. Well dusted stock.

2. Middlings which have been well graded.

3. Regular and constant feed.

4. Unvarying speed.

5. Large steady running fan and the machine air-tight.

6. Large room for expansion inside the purifier.

7. The rooms where purifiers are at work should be well ventilated and a constant supply of fresh air available, instead of the continued return of the same for successive duty, because in the case of the latter condition of things the air gets stale and moist and the temperature is increased beyond its natural surroundings, which leads to further condensation and consequent mischief.

8. The divisions, or sections, of each machine should be perfectly independent of each other; there should be a separate valve for each section, so as to regulate to a nicety the amount of air requisite for that particular locality and condition of feed.

9. Cut-off worms should be on all purifiers to allow of any part of the throughs or middlings being diverted to any other machine that may be necessary or convenient, apart from the bulk of the product.

10. Plenty of aspiration on the overtails, to correct or further help the efficiency of the general machine.

11. The finer the stock the greater should be the slope of the sieve by the aid of the suspenders and thumbscrews.

12. Judicious arrangement of covers to enable the greatest quantity of work to be accomplished with the least travel of feed.

13. If more impurities are noted in one place than in another, the feed is thinner there, and requires a readjustment of the sieve.

The break flour being brown and of less value than is usually obtained—the wheat, same mixture and seemingly clean—what would you suggest as the cause of this difference ?

If the wheats were washed and dried, I should suspect they had been dried too much, and the bran was too brittle and not tough enough. Or, I should suspect there was too light a feed on the mill and it was dressing up too bare. Or if fresh sharp rolls had been put in any of these breaks, especially with those dealing with the cleaning of the bran, I should suspect this to be the cause, or it might be that these breaks were too close. Or it might be that the scouring and brushing machines were not doing their work efficiently, especially if they were dealing with unwashed wheat which was intended to feed the mill; or it might be the difference in the way the break meal dressing machines were doing their work, on account of the atmosphere being dryer; or on account of the mill just having started but a short time, and all the machines being cold; or it might be that some new covers had been put on, which, although of the same numbering, would certainly dress freer.

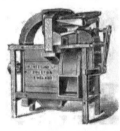

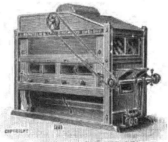

THE
MORRIS
AUTOMATIC SPRINKLER.

—— THE ——
IDEAL SPRINKLER
—— FOR ——
FLOUR MILLS.

SIMPLE

IN

CONSTRUCTION.

CERTAIN

IN

ACTION.

APPROVED FOR HIGHEST DISCOUNTS BY
ALL INSURANCE COMPANIES.

NO CLOGGING OF THE FUSIBLE JOINT BY STIVE.

ADOPTED BY LEADING MILLERS.

ESTIMATES AND ALL PARTICULARS—

THE SPRINKLER CO., LIMITED,

ST. BENE'T CHAMBERS,

1, FENCHURCH STREET, LONDON, E.C.

Telegrams—"WATERCRAFT, LONDON."   Telephone—2091 AVENUE.

Robinson's System.

The Best of Everything for Millers.

The most Scientific System on the market.
Produces the highest percentage of Pure
Flour. Contains the most efficient Purifier;
the most resultful Roller Mill; and the
latest Labour-saving Devices.

Thomas Robinson & Son,

Ltd.,

Manufacturers of Milling Machinery.
Flour Mills and Silos of any capacity
erected and equipped with the latest
and most successful appliances.

Head Office and Works: Rochdale, England.
London Office and Showrooms: 59, Mark Lane, and
1, Seething Lane, E.C.
Sydney (N.S.W.) Office & Depot: 317 & 319, Kent Street.
Cape Town: S. Walsh, Milling Expert, 370 G.P.O.

Agencies in :—

Amsterdam.	Gothenburg.	Durban.
Antwerp.	Ekaterinburg	Bombay.
Liége.	(Siberia)	Calcutta.
Barcelona.	Constantinople.	Shanghai.
Valletta.	Smyrna.	Yokohama.
Bale.	Port Elizabeth.	Manchuria.
Warsaw.	East London.	Mexico.
Moscow.	Johannesburg.	Buenos Ayres.

CPSIA information can be obtained
at www.ICGtesting.com
Printed in the USA
BVHW041650010920
587783BV00003B/190

9 781375 613941